## ALSO BY JOHN ELDER

*Intro To Python Programming*

*Ruby Programming For Everyone*

*Learn Ruby On Rails For Web Development*

*PHP Programming For Affiliate Marketers*

*The Smart Startup: How To Crush It*
*Without Falling Into The Venture Capital Trap*

*Adsense Niche Sites Unleashed*

*Social Media Marketing Unleashed*

*SEO Optimization: A How To SEO Guide*
*To Dominating The Search Engines*

# The AI Operating System

John Elder

SectionZero.AI

Las Vegas

# The AI Operating System

By: John Elder

Published By Narrow Canyon Publishing

Las Vegas, NV USA

ISBN

979-8-9989061-2-1

First Edition



http://SectionZero.AI

FOR APRIL

# TABLE OF CONTENTS

# The AI Operating System

Automate Almost Any Task in Your Work and Life Without Writing a Single Line of Code

# ABOUT THE AUTHOR

John Elder is a Web Developer, Entrepreneur, and Author living in Las Vegas, NV. He created one of the earliest online advertising networks in the late nineties and sold it to publicly traded WebQuest International Inc. at the height of the first dot-com boom.

He went on to develop one of the Internet's first Search Engine Optimization tools, the Submission-Spider that was used by over three million individuals, small businesses, and governments in over forty-two countries.

These days John writes about Coding and wealth creation; produces a popular coding Youtube Channel, and runs **Codemy.com** the online learning platform that teaches Coding to millions of students. Most recently he founded **SectionZero.AI** where he teaches cutting edge AI to a growing community of members.

John graduated with honors from Washington University in St. Louis with a degree in Economics. He can be reached at john@SectionZero.ai

# INTRODUCTION

AI companies like OpenAI and Anthropic need you to believe something and they really need you to believe it... They need you to believe that AI is really *really* complicated. Almost magically complicated.

Why? Because they've spent hundreds of billions of dollars on data centers with ungodly expensive AI chips to build these large language models (LLM's); and to make that money back, they have to charge you more and more for the tokens you use to run them.

And it's a lot easier to charge a fortune for something you don't understand. Mystery is profitable. Complicated means pricey.

The popular tools need you to believe it too. Take OpenClaw and Hermes Agent. These are open source, meaning anyone can read the source code and see exactly how they work. But they're betting you never will. They're betting you see them as mysterious black boxes. You fire up OpenClaw, give it some instructions in plain English, and it magically does what you asked. Token price be damned.

Except token price should not be damned. Tokens get more expensive all the time, and the bills add up fast.

I have a friend who started using OpenClaw without really knowing what he was doing. He fed it a pile of instructions on a Friday and went off to enjoy his weekend. When he checked on it Monday morning, it had quietly burned through $1,100 in API charges, churning tokens from ChatGPT and Claude the entire time he was gone.

CEOs need you to believe AI is complicated too. Otherwise it looks bad when they lay off another 10,000 people and claim they've "replaced them with AI." Pull out the magnifying glass and you'll find that very few of those companies are actually using AI to do that work yet. It just sounds good in a press release, and it nudges the stock price back up.

Consider Uber. Uber did what looked like the responsible thing. They gave their entire engineering team access to Claude and set up what they assumed was a nearly unlimited token budget for 2026. By April, the engineers had blown through the whole thing. Tens of millions of dollars, maybe more, gone in three or four months.

And what did they have to show for it? New features? New product lines? A measurable jump in engineering output?

Nothing.

But hey, AI is complicated, so it's not their fault.

We won't even get into the army of freshly minted "AI consultants" swarming small and medium businesses, charging astronomical fees to bolt AI onto companies that don't need them. You can guess what those consultants need you to believe.

But...what if AI wasn't super complicated?

What if you could use it simply, cheaply, and effectively without needing to learn any coding or super secret AI techniques?

What if you could just sort of talk your way through what you needed to automate, and the AI itself could create a simple workflow on your computer...without any fancy expensive third party tools...

And what if it all just...worked?

That's what this book is here to teach you.

Because AI isn't super complicated. Once you understand a few very basic concepts, you can do more with AI than those over-hyped CEO's, the high priced AI consultants, the OpenClaw fanboys, and really - everyone.

And you can do it without ever worrying about waking up to a surprise $1,100 API bill.

So buckle up. This is gonna be a lot of fun and it could honestly change your life. Because once you truly understand AI...there's not much you can't do. We're at a rare moment in history where the people who understand this stuff and actually take action stand to make an awful lot of money, very quickly.

# PART I: WHY THIS, WHY NOW

# CHAPTER ONE

## You're Not Too Late (and You're Not Getting Replaced)

It's 11:40 on a Tuesday night and you're doing the thing you promised yourself you'd stop doing. You're in bed, phone six inches from your face, reading about AI.

Another company announced layoffs today. The press release used the word "efficiency" four times and "AI" seven. Somewhere in the article an executive said the future belongs to companies that "embrace the transformation." You read the comments. Half of them say this is the end of work as we know it. The other half say it's all hype. Neither half makes you feel any better.

If you're an employee, the question under your ribs is simple: how long until that's my company, my department, my desk?

If you run a small business, your question is different but it lives in the same spot: everyone keeps telling me I should be "using AI," and I have no idea what that actually means. Meanwhile I keep hearing about some competitor who's doing... something... with it.

Two different people. Same knot.

I wrote this book for both of you, and I want to start by telling you something that the layoff headlines, the LinkedIn gurus, and the doom-scrolling comment sections all get wrong.

You're not too late. And you're not getting replaced.

Not because AI is overhyped. It isn't. It's the most useful new tool I've encountered in all my years of building businesses. You're not getting replaced because of something more specific, and once you see it, you can't unsee it:

AI doesn't take jobs. People directing AI take tasks.

That sentence is doing a lot of work, so let me unpack it slowly.

### The machine has no wants

Here's a fact so obvious that almost everyone forgets it: AI doesn't want your job.

AI doesn't want anything. It has no goals, no ambition, no Monday mornings. Every AI system in existence sits completely idle until a human being tells it to do something. It doesn't wake up early to impress the boss, and it doesn't network. The most powerful model on earth, left alone, does exactly nothing, forever.

Which means every scary AI story you've ever read is actually a story about a person. A person decided what the AI would do. A person pointed it at a task. A person checked the output, or didn't. When a company "replaces workers with AI," what actually happened is that some human beings figured out how to direct AI at tasks that other human beings used to do by hand.

I'm not telling you this to make you feel better. I'm telling you because it changes where the threat actually lives, and where the opportunity does.

The threat was never the machine. The threat, if there is one, is a person who knows how to direct the machine, competing against a person who doesn't.

And the opportunity is that becoming the first kind of person is dramatically easier than anyone is telling you.

## The divide that actually matters

Every big workplace technology has split people into two groups. Not the skilled and the unskilled. Not the young and the old. The split is always the same: the people who picked the tool up, and the people who waited.

When the spreadsheet arrived, the accountants and analysts who learned it didn't disappear. They became the people who could answer in an afternoon questions that used to take a week of pencil work. The ones who waited found that the patience for pencil work was evaporating around them.

When word processors arrived, when email arrived, when the web arrived, same story. The tool didn't decide anyone's fate. The waiting did.

I know this argument has been used to sell every gadget of the last forty years, so let me be honest about where AI fits it and where it doesn't.

Here's where it fits: AI is a tool that multiplies the person using it, exactly like the spreadsheet multiplied the accountant. The multiplication is real. Later in this book I'll show you a task that used to cost my business about ninety dollars and five hours of skilled human work, every single time it ran. It now runs in about a minute and costs a nickel. Nothing about my talent changed between those two numbers. Only the directing did.

Here's where it doesn't fit, and where I understand the fear: no spreadsheet ever wrote you a paragraph. AI talks. It answers in full sentences, it apologizes, it explains itself. It feels less like a tool and more like a rival, a very fast intern with infinite stamina. When something talks like a person, we instinctively rank ourselves against it.

Don't. Ranking yourself against AI is like ranking yourself against a forklift. The forklift lifts more than you. So what? The only interesting question is who's driving it.

## Two Mondays

Since this book has two readers, let me walk a Monday for each of you, because the same divide shows up in both lives wearing different clothes.

Here's the employee's Monday. You're a marketing coordinator, or an office manager, or a paralegal, and you spend the first ninety minutes of every week on the same ritual: pulling numbers from three systems into one report, reformatting what someone sent you Friday, updating the tracker nobody reads until they urgently need it. You're good at your actual job, the judgment parts, the people parts, but the ritual eats the morning, and lately there's a new flavor to it: a suspicion that this is exactly the stuff that's going to get you replaced. Here's the reframe I want you to try on. That ritual is not your job. It's the barnacle on your job, and you know it better than anyone alive, every exception, every weird Thursday in the data, every reason the obvious shortcut doesn't work. That knowledge makes you the only person who can actually automate it well. The coordinator who shows up having turned the ninety-minute ritual into a nine-minute review isn't the one who gets replaced. She's the one who gets asked how she did it, and "I can show you" is the strongest sentence a career has ever been built on.

Now the owner's Monday. You run a landscaping company, or a dental practice, or an online store, and your Monday isn't ninety minutes of ritual; it's a hailstorm of them. Invoices, scheduling, the supplier's new price list, the review someone left Saturday, the payroll question, the quote that needs to go out. You've heard you should be "using AI" so many times the phrase has gone blurry, and every time you've looked into it, you found either a chatbot that forgets you or an enterprise platform priced for a company with an IT department. Here's your reframe. You don't have an AI problem; you have a task-inventory problem. Somewhere in that hailstorm are five or six tasks that recur, follow rules, and don't need your judgment, and each one is a workflow waiting for the twenty-minute conversation you'll learn in Chapter 10. Nobody sells this to you because there's nothing to sell: no seat licenses, no implementation fees, no consultant. Just your folder, your tasks, and a method small enough for a business of one.

Different Mondays, same move: find the described-able, repeated, judgment-light work, and hand it over. The employee's version protects a career. The owner's version buys back the evenings. The method, you'll notice for the rest of this book, is identical.

### What "directing" actually means

At this point you might expect me to tell you to learn "prompt engineering," or take a course in Python, or get some certification with AI in the name.

No.

The skill this book teaches is direction, and I want to define it precisely, because the word sounds vaguer than it is. Directing AI means three things:

First, describing work. Taking a task you understand and putting it into plain words: what comes in, what should go out, what matters along the way. If you've ever written instructions for a house sitter, you've done this.

Second, giving the work a home. Setting up a simple structure of files and folders where the AI operates. This sounds technical. It's the opposite. If you've ever organized a filing cabinet, you've done this too. We'll spend a good chunk of this book on it, and you'll be surprised how simple it is.

Third, course-correcting. Looking at what came back, noticing it's not quite right, and saying so in plain English. "This report is too long. Cut it in half and lead with the numbers." That's course-correcting. You've been doing it your whole working life; you just did it to people instead of software.

Let me show you what those three skills sound like in the wild, because "directing AI" stays abstract until you hear it. Here's a condensed version of a real exchange, the kind you'll be having by Chapter 10. The human says: "Every week I get emailed a spreadsheet of orders. I need the rush orders pulled out and turned into a list for my warehouse guy: item, quantity, deadline, nothing else. Rush means ships within three days." The AI asks: "What should happen with orders marked rush that are missing a ship date?" The human, who has just realized he never thought about that in nine years of doing this by hand, says: "Flag those separately, actually, those are the ones that burn us." Three sentences of description, one good question, one course-correction, and notice what never came up: code, software names, anything technical at all. That's the entire dialect. If you can read that exchange and imagine holding up your end of it, and you can, you already speak the language this book is written in.

Notice what's not on the list. Writing code isn't on the list. Reading code isn't on the list. Understanding how large language models work isn't on the list. In the system I'm going to teach you, the AI writes whatever code is needed, runs it, and fixes it, out of your sight. You will direct work you never touch.

If you've ever trained a new employee, delegated to an assistant, or explained to a teenager exactly what "clean your room" means, you already have the raw skills. What you're missing is a method. That's what this book is for.

## The moment it flipped for me

Let me tell you about the ninety-dollar task, because it's the moment this stopped being theory in my own business.

I run YouTube channels, a lot of them, and it's been my full-time business for years. One of the things that makes the channels work is that every video gets its own dedicated page on the channel's website. Not a lazy embed. A real page: an article, a formatted transcript, a set of questions and answers, all built around the video, all tuned so that search engines and AI answer engines can find it.

For years, a person built those pages. It took about five hours per page, and I paid eighteen dollars an hour for careful, skilled work. Five pages a week. Do the math and it's roughly eighteen hundred dollars a month, month after month, year after year. It was worth every penny, because the pages bring in traffic, and traffic is the business.

Then I built what you're going to learn to build in this book. Today, when a video goes live, I hand my AI the video's web address. A small program fetches the video and transcript. The AI writes the article and the Q&A, formats the transcript, assembles the page, and publishes it to my server. One instruction from me. About sixty seconds of running time. The bill comes to a few cents.

I want to be straight with you about both halves of that story, because each half matters.

The first half: no programmer was involved. I described the task the way I just described it to you, in sentences. The AI and I worked out the steps together, it wrote the small programs, and when something came out wrong, I told it what was wrong and it fixed it. The whole thing was built in conversation.

The second half: a real person used to do that work, and now they don't. I won't pretend otherwise, and I won't pretend that's painless. This is the fear from the beginning of the chapter, made concrete. But look carefully at what actually happened. The task didn't vanish because AI wanted it. It moved because someone who understood the work, me, directed AI at it. That's the whole event. The decision, the knowledge of what a good page looks like, the judgment about what to automate, all of it stayed human. It just concentrated in the person holding the directions.

That's the divide, up close. It has nothing to do with whether you can code. The person I paid could not have been protected by learning to code. And I wasn't empowered by coding; I was empowered by knowing the work cold and having a method for handing it over.

You know some piece of work cold. Everybody does. Your inbox, your invoices, your inventory, your client reports, your listing photos, your weekly numbers. That knowledge is the scarce ingredient. The method is the cheap part, and you're holding it.

## What I'm promising you

Books like this one tend to overpromise, so let me put my cards on the table. Here's what will be true by the time you finish.

You will understand a complete system for automating real tasks. It's called the AI Operating System, and despite the name, it's not software you install. It's a way of organizing work: one folder on your computer that holds your automations, your data, and your instructions, with an AI directing traffic inside it. It costs almost nothing to run. You'll understand every piece of it, not because you'll read the code, but because every piece has a plain-English job you can name.

You will build a working automation of your own, from scratch, by having a conversation. Not a toy. A real workflow that does a real task you currently do by hand, and that you can run again any time you want.

And you will know how to keep going: how to add the next workflow, and the one after that, until the dull third of your working week belongs to the machine.

Now the other side of the ledger, because trust is built with both columns.

This is not magic. When a workflow misbehaves, and eventually one will, you'll fix it by describing the problem to your AI and going back and forth, the same way you'd sort out a mix-up with a contractor. That back-and-forth is not a failure of the system; it is the system. You're not going to be coding, but you are going to be engaged. If you want a machine you never have to talk to, that machine doesn't exist at any price.

And this is not instant. Your first workflow will take you an afternoon, most of it spent learning the rhythm of the thing. The tenth will take you twenty minutes. The payoff compounds; the first step is honest work.

That's the deal. No code, no consultants, no monthly platform fees, no surprise bills. Just you, a folder, and a method.

One more promise-shaped thing, about timing, since "you're not too late" is half this chapter's title. There's a myth forming that the AI advantage went to whoever got in earliest, the way people talk about early Bitcoin or early Instagram. It's backwards here. The people who started two years before you had to fight tools that were genuinely hard: clumsy interfaces, models that lost the thread, workarounds on workarounds. You're arriving exactly when the tools crossed the threshold this book depends on, when an AI can hold a long conversation, work inside your folders, and write its own code reliably enough to trust with real tasks. Being "late" to AI in this moment is like being late to email in 1997: the party, for normal people with normal businesses, is just starting, and the door is propped open. The advantage that matters isn't a head start measured in years. It's the method, and method is exactly what the early adopters mostly didn't have. You're holding it.

## Start noticing

I want to leave you with an assignment, and I promise it's the easiest one in the book. You don't need a computer. You don't need an AI. You need a scrap of paper.

For the next few days, just notice. Notice every time you do a task you've done before, the same way, more than twice. The report you assemble every Friday from the same three places. The invoices you rename and file. The competitor's website you check each morning. The photos you resize before every listing. The same eleven questions answered in the same polite email, again and again.

When you catch one, write down one sentence about it: what goes in, what comes out. "Every Monday I collect last week's sales numbers and email a summary to my partner." That's the sentence. That's the whole exercise.

To prime your noticing, here's what these sentences look like when other people collect them. A realtor: "Every new listing, I resize the photos, write the description, and post it to three sites." A restaurant owner: "Every morning I check four suppliers' sites for price changes on my top twenty ingredients." A therapist in private practice: "After every session I write an insurance-formatted note from my shorthand." A freelance designer: "Every project, I assemble the same contract, invoice, and kickoff email with new names and dates." A property manager: "Every rent week, I match payments against the tenant list and chase the stragglers." None of these people think of themselves as technical, every one of those sentences is a workflow this book can build, and the pattern to steal is the shape: in, out, repeat.

Most people, most weeks, collect five or six sentences without trying. Keep the list. It's not busywork; it's the raw material for everything that follows. By the end of this book, the items on that list will be workflows in a folder, and the list itself will have taught you something worth more than any single automation: the habit of seeing your own work as a set of tasks that can be described.

Because here's the quiet truth underneath this whole chapter. The person who can describe a task clearly has always been the person in charge. Foremen, editors, executives, generals: their tool was description. AI didn't change that. AI just made description executable.

You've been qualifying for this job your whole life. Let's go do it.

# CHAPTER TWO

## What AI Automation Really Is

You've already used AI. You typed a question into a chat window, watched an answer assemble itself word by word, thought "huh, that's pretty good," and closed the tab.

That experience is nearly universal now, and it has quietly installed a picture in millions of heads: AI is a thing you talk to. You ask, it answers, the conversation ends, and everything it said evaporates. Tomorrow you'll introduce yourself all over again.

Useful? Sure. I use chat windows every day. But a chat window is to AI automation what a phone call is to a shipping company. The call matters. The call is not the trucks.

This chapter is about the trucks. Before I can teach you the system, we have to agree on what "AI automation" even means, because the phrase has been stretched to cover everything from a toaster with a chip in it to science fiction. Most of what you've heard is wrong in one particular direction, and the wrongness is expensive. So let's build the real picture from the ground up, starting with a word you'll see on nearly every page of this book.

### What a workflow is

Strip away the technology entirely. A workflow is just a task broken into steps that run the same way every time.

You are surrounded by them. Payroll is a workflow: collect the hours, check them against the schedule, calculate the pay, send it, file the record. Closing a restaurant kitchen is a workflow: strip the stations, store the food, scrub the surfaces, check the temps, lock the walk-in. A preflight checklist is a workflow so serious that pilots aren't allowed to skip it even when they're certain.

Notice three things about every workflow on that list.

The steps are knowable. You could write them on a card and hand the card to a competent stranger, and the job would get done.

The order matters. You don't scrub the station and then strip it.

And almost none of it requires judgment. Real judgment shows up in flashes: is this fish still good, does this pay stub look off, why is the fuel gauge reading low? The rest is faithful execution. Any honest look at any recurring task finds the same ratio: a lot of doing, a little deciding.

Hold onto that ratio. It's about to become the most important idea in this book.

An automated workflow is nothing more exotic than this: the card you'd hand the stranger, handed to your computer instead. The steps get written down in a form a machine can follow, and the machine follows them. When people say "automation," that's the whole trick, and it's been the whole trick since long before AI: written-down steps, faithfully executed.

One more thing about workflows before the technology enters: you run more of them than you think, because the human mind is talented at making repetition invisible. Nobody experiences their Tuesday as "executing workflows"; they experience it as "being busy." The office manager doesn't think of the supply order as a seven-step procedure; she just orders supplies, the way she always does, which is exactly the point. "The way I always do it" is the civilian phrase for a workflow. Any time you hear yourself say it, or its cousins, "let me just knock this out," "the usual routine," "I could do this in my sleep", your brain is telling you it long ago wrote the card and handed it to itself. This book's whole opening move is getting those cards out of your head and onto paper, which is why the Chapter 1 exercise asked for sentences, not software.

So where does AI come in? Ah. That's where everyone goes wrong.

## The picture in everyone's head

Ask a smart, busy person what "AI automation" looks like, and they'll describe something like this: you get access to a powerful AI, you tell it "handle my invoices," and it... handles your invoices. The model does everything. It reads, it decides, it files, it remembers, it acts. The AI is the whole machine.

I understand why this picture is in everyone's head. It's the picture the demos sell. It's the picture the agent tools we met in the introduction are built on. It's the picture behind every breathless post about someone's AI "running their whole business." The model is the star, so the model gets every job.

Here's my problem with the picture, and it's not a small one. The model is the most expensive, slowest, and least predictable component in the entire toolbox. Handing it everything is like hiring a brilliant, pricey consultant and asking her to also carry the boxes, staple the receipts, and stand in line at the post office. She can do all that. You will get a staggeringly expensive invoice, and some days the receipts will be stapled in an order that made sense to her at the time.

I'll spend all of Chapter 3 on why this matters, with real dollar figures that should make your eyebrows move. For now I'll just plant the flag: in this book, the AI does as little as possible. Not because AI is bad. Because AI is precious, and you don't spend the precious thing on work a dumb, free, tireless machine can do perfectly.

While we're clearing the picture in everyone's head, a word about where it comes from, because knowing the source inoculates you. Nearly everything you've seen about AI automation arrived as a demo: a two-minute clip of an agent doing something jaw-dropping, booking a trip, building an app, running a store. Demos are real, mostly, but they're also survivors: you see the take where it worked, not the forty where it wandered, and you see minute two, never month two. A demo has no memory to maintain, no Tuesday-after-Tuesday consistency to keep, no bill arriving at the end. The picture this book is giving you is the opposite of a demo. It's boring on purpose: a system optimized for the four-hundredth run, not the first one. When you've finished this book, you'll watch those clips differently, and the question you'll ask is the one nobody asks in the comments: "What does run four hundred look like, and what did it cost?"

Which raises the obvious question: if the AI isn't doing everything, what does the automation actually look like?

## A conversation and a folder

Every automation you build in this book comes down to two things: a conversation and a folder. That's the method. Let me introduce both.

The conversation happens in an AI application running on your computer. As I write this, I mainly use two: Anthropic's Claude desktop app and OpenAI's Codex app. Throughout this book I'll show examples in both, and I want to say this clearly, right here at the first mention, because it's a core belief of the whole system: the specific app does not matter. These two happen to be the strongest as I write this sentence. By the time you read it, there may be a third, or the two may have traded places. Any AI tool of this general caliber, one that can hold a conversation, read and write files in a folder you point it at, and write and run small programs, will work. The method you're learning is bigger than any logo, and in Chapter 8 I'll show you how completely swappable these tools are, because that swappability is a feature you're going to care about.

What I will not be showing you: a terminal. If you've seen intimidating videos of programmers typing green commands into black windows, exhale. These apps we'll use look like a chat window, because that's what they are. You type sentences. It types back. The difference between this and the chat window you've already used is everything that happens around the conversation: this one has its hands.

If you're wondering how to choose between the two apps, here's the honest answer: for everything this book teaches, it barely matters, and I'd rather you spend five minutes deciding than five days researching. Both let you point at a folder on your own computer and talk. Both write and run their own code invisibly. Both handle the workflows in this book without breaking a sweat. The models behind them each have a personality, one may write a little warmer, the other a little tighter, and you'll develop a preference the way you develop a preference between two good coffee shops: by going a few times, not by reading reviews. Pick whichever you've heard of, or whichever a friend uses, and start. Chapter 8 will show you that this choice is nearly weightless, because in this system, switching later costs a few minutes, not a lengthy rebuild.

Which brings us to the folder.

When you set up an automation, you point the AI app at an ordinary folder on your computer and you have a conversation about a task. Together, and I mean that literally, you and the AI fill that folder with a few plain files: a file of written instructions describing the task, a small program or two that the AI writes to do the mechanical steps, a place to keep records, and a log that notes what happened on each run. The folder is the automation. It sits there, visible, ordinary, no different from a folder of vacation photos. You can open it. You can back it up. You can drag it to another computer.

I'll walk you through every file in Part IV, and you'll build your own from scratch in Chapter 10. Right now I just want the shape in your head: conversation in, folder out. No installation wizard, no monthly platform, no dashboard with seventy toggles. A talk, and a place where the results of the talk live.

Let me make it concrete.

## Watching one get born

Meet Dana. I invented her for this example, but her problem is drawn from life; you'll meet my real workflows, with real dollar figures, in Chapter 9.

Dana runs a photography studio. Every Monday morning she spends about forty minutes assembling her week: she opens her booking system and downloads the CSV file of the coming week's sessions (a CSV is just a spreadsheet saved as plain text; her booking system spits them out). She cross-checks it against her notes for special requests. Then she writes a summary for her two part-time shooters: who, where, when, what gear.

Forty minutes, every Monday, forever. It's exactly the kind of sentence I asked you to collect at the end of Chapter 1: sessions come in, briefing goes out.

Here's Monday number one of Dana's new life. She opens her AI app, points it at a new folder she's named `weekly-briefing`, and types, more or less:

"Every Monday I download a CSV of the week's sessions from my booking system. I'll drop it in this folder. I need a briefing for my two shooters: each session's client name, location, start time, and gear notes. Gear notes live in the special-requests column. Group it by shooter. Write it so I can paste it straight into a text message. Can you set this up as a workflow I can run every week?"

The AI answers in plain English. It asks to see a sample CSV, and she drags one in. It asks two questions she hadn't thought to specify: what should happen with sessions that have no shooter assigned yet, and does she want a warning when two sessions overlap? (Yes, and yes. Good questions. Notice the AI is contributing, not just obeying; this is a conversation, not a command line.)

Then it says: here's my plan. I'll write a small program that reads the CSV and pulls out the fields you named, because that part is mechanical and shouldn't need AI at all. I'll keep your formatting preferences in an instructions file. When the program has extracted the week's data, I'll write the briefing in the tone you asked for and flag anything unusual. I'll log each run. Sound right?

Dana says yes. Ninety seconds later the folder contains a handful of files, and her first briefing is on the screen. It's good. The gear notes are grouped a little oddly, so she says "put rentals on their own line," and the AI adjusts the instructions file so it stays fixed. Total setup time: about twenty minutes.

Monday number two is the part I want you to see. Dana drops in the new CSV and types: "Run my weekly briefing." That's it. That's the whole interaction, this Monday and every Monday after. A minute later, there's her briefing. Her forty minutes now costs her about ninety seconds, most of which is the drag and the drop.

## The three workers in the folder

Rewind Dana's setup and look at who's actually doing what inside that folder, because the division of labor is the entire philosophy of this book in miniature.

The program (probably written in Python) does the drudgery. Opening the CSV, finding the columns, pulling the fields, checking for overlaps: a small program the AI wrote handles all of it, identically, every single week. This is the "faithful execution" part of the workflow, the many-steps-no-judgment part, and it runs for free (no AI tokens needed). Dana never sees this program. She couldn't read it if she opened it, and she never needs to open it.

The files do the remembering. The instructions file holds what Dana wants: the format, the tone, "rentals on their own line." The log holds what happened. Next Monday, nothing needs re-explaining, because the folder remembers. This is the difference between a chat window, which wakes up every morning with no idea who you are, and a workflow, which accumulates.

The AI does the judgment, and only the judgment. It reads what the program extracted and writes a briefing a human would want to read: smooth, warm, "heads up, Saturday is tight, two sessions forty minutes apart." That's real judgment work, the flash of deciding, and it's the only part of the weekly run that touches the AI at all.

A lot of doing, a little deciding. The kitchen ratio, the payroll ratio, now built in software: the doing pushed down into a free, tireless program, the remembering pushed into plain files, the deciding, and only the deciding, handed to the precious AI component.

And because Dana runs a photography studio and you probably don't, let's transpose her anatomy onto two other businesses, quickly, so you can see the skeleton through different skin.

A general contractor who assembles bids: the program reads his cost spreadsheets and the supplier price files, does every calculation, and totals the line items; the files remember his margins, his boilerplate, and every bid he's ever sent; the AI writes the two paragraphs that explain the approach in the client's language and flags anything that looks off against similar past jobs.

A clinic office manager who chases unpaid claims: the program pulls the aging report and sorts out which claims crossed thirty days; the files remember which insurers respond to what and the history of every claim; the AI drafts the follow-up letters, each one worded for its insurer, and routes the two weird cases to a human. Same three workers every time, doing, knowing, deciding, and by now you can probably run the transposition yourself for your own shop. That's not an accident. That's the book working.

There's a name for the AI's role here, and it's the name this whole book hangs on: the AI is the orchestrator. It doesn't play every instrument. It stands in front, decides what runs when, and shapes the result. You'll get the full picture in Chapter 6, including the one place the metaphor really pays off: if the AI is the conductor, you're the one holding the baton over the conductor. You direct the orchestrator. It directs the work.

And one more time, because it's the promise on this book's cover: nobody in Dana's story wrote code except the AI. Nobody read code, either. Dana's contributions were sentences: a description, two answers, one correction. Every skill she used, you had before you picked up this book.

## Chatting versus owning

So let's come back to the chat window, because now you can see exactly what it was missing. When you chat, there's no folder. Nothing persists, nothing accumulates, nothing runs while you're not there, and next time you start from zero. Chatting is renting an intelligence by the minute. A workflow is owning a system.

The difference shows up in three places. Memory: Dana's folder knows her shooters, her format, her preferences, and your chat window knows nothing about last Tuesday. Repeatability: "run my weekly briefing" produces the same shape of result every week, while the same request retyped into a chat produces whatever mood the model is in that day. And cost, which deserves its own chapter and is about to get one: in Dana's weekly run, the AI is awake for seconds, because the free python program did everything else first.

Play the difference forward a year and it stops being subtle. The chatter has had five hundred conversations, some of them wonderful, and owns nothing but a scroll of transcripts; every Monday still starts from zero, and the quality of any given week depends on how well he re-explains himself that morning. Dana has run one workflow fifty-two times. Her folder now holds a year of briefings, a memory of every session her studio has shot, and instructions that have absorbed a year of her corrections; her Monday costs ninety seconds, and this year's ninety seconds is better than last year's, because the system has been quietly compounding while the chatter has been quietly repeating. Same underlying AI, possibly the identical model. The entire difference is the folder.

That last one, we've barely touched. It's time. Because once you see what the all-AI picture actually costs, in dollars, in reliability, and in control, you'll understand why this book's whole method fits inside one contrarian sentence:

The less your AI does, the more it's worth.

That's Chapter 3.

# CHAPTER THREE

## Why Less AI Is More

In the spring of 2026, a developer named Peter Steinberger ran up a bill of $1.3 million.

Not for the year. For thirty days.

Steinberger is the creator of OpenClaw, which at that moment was the fastest-growing open-source project in the history of GitHub, the site where the world's programmers share code. He was running about a hundred AI agents at once, around the clock, writing code, reviewing code, hunting bugs. The agents did their thinking by calling OpenAI's models, and thinking is billed by the token. In one month his hundred agents consumed 603 billion tokens across 7.6 million requests. Price tag: $1.3 million.

Now, the honest caveats, because this book doesn't do gotchas. Steinberger works at OpenAI, and OpenAI covered the bill. The figure reflects premium "fast mode" pricing; at standard rates it would have been closer to $300,000 a month. And a hundred simultaneous agents is an extreme experiment, not a normal person's setup. Nobody reading this book is in danger of a seven-figure API bill.

But I'm not telling you the story for the number. I'm telling you for the ratio. Divide it out and even the charitable version comes to thousands of dollars per agent, per month. For software that's free.

Sit with that for a second, because it's the strangest pricing model you've ever encountered. OpenClaw costs nothing. Anyone can download it. The code is open for inspection. And running one instance of this free software at full tilt can cost more than a car payment. Every month.

Why? Because of a design decision. And understanding that one decision is worth more to you than everything else in this book's first hundred pages, because it's the decision you're going to make differently.

## The question nobody asks

Every automation, everywhere, is an answer to a question that almost nobody thinks to ask out loud:

How much of this task actually needs a brain?

Remember the kitchen ratio from Chapter 2: every recurring task is a lot of doing and a little deciding. The doing is mechanical. Fetch this, rename that, extract these columns, publish there. The deciding is the flash of judgment in the middle: what should the summary emphasize, is this result weird, which of these matter?

The tools like OpenClaw making headlines answer the question one way: all of it. Give the brain everything. Every fetch, every rename, every step routed through the model, reasoned about in tokens, billed by the syllable. The brain carries the boxes.

This book answers the question the other way: almost none of it. Push every mechanical step down into ordinary code that runs for free, store everything worth remembering in plain files, and wake the AI only for the moments that need actual judgment.

Same task. Same result. The difference is where the work lives, and that difference shows up in three places that matter to you: your wallet, your sleep, and your control. Let's take them in order. But first, let me properly introduce the tools on the other side of the argument, because you're going to keep hearing about them, and because they're better than my snark in the introduction suggested.

## A fair look at the all-brain machines

Two tools define the "give the brain everything" school as I write this: OpenClaw and Hermes Agent. Both are free and open source. Both are impressive. I mean that without irony.

OpenClaw is Steinberger's creation, first released in late 2025 (it burned through two earlier names, Warelay and Moltbot, before landing on OpenClaw; the lobster theme is a long story). It's a personal AI agent that lives inside your messaging apps. You text it on WhatsApp or Telegram or Slack, like texting an assistant, and it goes off and does things: manages your calendar, browses the web, runs errands across your accounts, files your email. It learns new tricks through "skills," little instruction packets the community writes and shares; thousands of them exist. Within months of release it had over 300,000 stars on GitHub, passing projects that had been accumulating fans for a decade. People love it, and I understand why. There's real magic in texting your computer from the beach.

Hermes Agent, from an outfit called Nous Research, arrived in early 2026 and pushed the idea further. It runs continuously on your own machine, remembers everything across sessions, runs scheduled tasks while you sleep, and, its signature move, watches for tasks you do repeatedly and writes its own skills for them. An assistant that grows itself. It works with almost any AI model behind the scenes, and it collected 175,000 GitHub stars in under four months.

I want to be fair to these tools, so let me say plainly: for what they're for, they're wonderful. If you want a conversational companion that can reach into your digital life from any chat app, they deliver, and the open-source communities around them are doing creative, generous work.

And there are people for whom they're the right choice. If you're a developer who wants a tinkerable agent and can read the code when it misbehaves, these projects are a playground. If your use is light and conversational, a question from the grocery store, a calendar shuffle, the token bills stay small, because small uses meter small. If an employer is paying, well, Steinberger's employer paid his. My argument was never that these tools are bad or that their users are fools; the star counts on GitHub include a lot of clever people. My argument is about fit: a business owner automating recurring work, the person this book is for, has the one profile these architectures punish. Your tasks repeat, so per-run costs multiply by fifty-two. Your tasks must come out the same each time, so improvisation is a bug. And nobody is covering your bill. For you, the all-brain design isn't wrong because it's dumb. It's wrong because it's aimed at someone else.

But notice the architecture, because it's the introduction's warning made flesh. Neither tool contains any intelligence of its own. Every step, understanding your text, planning what to do, deciding which skill to use, checking the result, reasoning about what's next, happens by calling an AI model and paying for tokens. The agent is, in a real sense, a beautifully engineered pipe between your chat apps and a rented brain. When it works on something for an hour, that's an hour of metered thinking. When it runs all weekend, the meter runs all weekend.

That's not a flaw. It's the design. These tools chose maximum brain because maximum brain gives maximum flexibility: an agent that reasons about every step can attempt almost anything. My friend from the introduction, the one who came home to an $1,100 weekend, wasn't misusing OpenClaw. He was using it exactly as designed. He just didn't understand what the design meant for his wallet.

You, on the other hand, are going to understand exactly what designs mean for wallets. Starting now.

### Reason one: the meter

A token, since I've used the word a dozen times, is just the unit AI companies bill in: a chunk of text a few characters long, roughly three-quarters of a word. Every word the model reads costs tokens. Every word it thinks costs tokens; when a model reasons through a problem step by step, all that internal deliberation is billed text too. Every word it writes costs tokens. The prices per token look microscopic until you notice how many tokens serious work consumes. Steinberger's 603 billion tokens in a month is the proof.

Here's how the billing actually feels in practice, because the microscopic prices are exactly what fools people. A model's rates are quoted per million tokens, in figures that sound like nothing: a few dollars, sometimes less. The trap is in what multiplies. An agent working on a task doesn't read your request once; it re-reads its whole growing pile of context, the instructions, the conversation so far, the results of every step, over and over as it works, because each new step needs the full picture again. So a task that involves twenty steps doesn't cost twenty small sips; it costs twenty increasingly large gulps, and a long transcript or a fat document sitting in the middle of the pile gets re-swallowed every single time. This is how my friend's quiet weekend became eleven hundred dollars: not one big purchase, but tens of thousands of unremarkable ones, compounding while nobody watched. Token billing is honest, metered, and utterly without mercy for architectures that think out loud all day.

Ordinary code has no meter. This is so important and so under-said that I'm going to say it twice. A program, once written (in say Python or Javascript or whatever programming language needed), costs nothing to run. Nothing meaningful, anyway; the electricity is rounding error. A program can extract columns from a spreadsheet a million times for free. It never gets tired, never charges overtime, and never bills you for reasoning, because it doesn't reason. It executes.

So take one task and run it through both architectures. The task: my video pages from Chapter 1. A video goes live; a full web page needs to exist, article, formatted transcript, Q&A section, published to my server.

Here's the all-brain version: an agent reads my request, reasons about it, browses to the video, reasons about the page, works through the transcript chunk by chunk (every chunk in, every chunk out, billed both directions), reasons some more, drafts, evaluates, formats, reasons about publishing, publishes. Tokens at every joint. Depending on the model and the day, that's a few dollars per page, and I publish five pages a week, every week. Call it fifty to a hundred dollars a month, if nothing goes weird. More if something does, because agents that hit a snag reason harder, and reasoning is the thing on the meter.

Here's my program-driven version: a small python program fetches the video data and transcript. Free. The AI wakes up once, with everything already on its desk, and does the one part that needs a brain: write the article, write the Q&A, shape the transcript. A single well-fed model call. The program assembles and publishes the page. Free. Total cost per page: three to five cents.

Three to five cents versus a few dollars is a factor of about a hundred, and it's not because my AI is worse. It's often the same model. The difference is that in my system, the brain never carries boxes. It arrives, judges, then leaves.

A hundred-fold difference doesn't matter much on one page. Across five pages a week for years, across the dozen other workflows in my folder, across everything you'll eventually automate, it's the difference between "my automations cost less per month than a sandwich" and "my automations have a line item in the budget." Uber, from the introduction, learned this at corporate scale: thousands of engineers, each with a metered brain running all day, and a year's budget gone by April. The meter doesn't care how big you are. It only cares how often the brain runs.

### Reason two: the 3 a.m. failure

Money is the argument everyone hears first, but it's not the one that converted me. This one is.

Ordinary code is deterministic. Feed a program the same input and it does the same thing, every time, forever. AI models are the opposite: probabilistic, which is a technical way of saying that the same request can produce different answers on different days. Usually the differences are cosmetic. Sometimes they aren't.

Now, everything breaks eventually. Automations too. The question that separates good systems from bad ones isn't whether they fail. It's whether you can find the failure.

When my page workflow breaks, it breaks like a machine. The video site changed its page layout, and the little fetch program came back empty. It came back empty at step two, the log says so, the log has the timestamp and the error, and the fix is one conversation with my AI: "the fetch step is failing, here's the log, figure out what changed and fix it." Machine failures are loud, located, and repeatable, which makes them cheap to fix.

When an all-brain agent breaks, it often breaks like a mood. Somewhere in a long chain of reasoning, the model interpreted a step a little differently than yesterday. No error fired, because nothing crashed; the agent confidently did a slightly wrong thing and sailed on. The failure surfaces later, downstream, as a report with the wrong week's numbers or an email sent to a list it shouldn't have gone to, and when you try to reproduce it, the agent, being probabilistic, does it right this time. Nothing to grab. Nothing to fix. Just a system you trust a little less than you did last month.

There's a rule of thumb hiding in this: the more of a workflow you hand to the model, the more of your failures become moods instead of machines. Flip it around and you get this book's design principle. In my workflows, the deterministic steps are a scaffold of certainty around the one probabilistic moment. If the input arrived and the log shows every step green, then whatever's off can only live in one place: the writing step. I never wonder where to look. There's only one room in the house where the lights can flicker.

There's a second cost hiding inside the mood failures, one that never shows up on any bill: the verification tax. Once a system has confidently done a slightly wrong thing to you twice, you start checking everything, and now you're spending the exact minutes the automation was supposed to buy back, plus a little anxiety on top. I've watched people quietly abandon impressive agent setups for precisely this reason, not because the failures were frequent, but because they were unlocatable, and unlocatable failures make every success feel provisional. An automation you have to audit every time isn't an automation. It's a coworker you don't trust, with better typing speed.

You can't read code, and you never will read the code in your folder. So this property matters even more for you than for a programmer. Your trust in the system won't come from inspecting it; it'll come from the fact that nearly all of it behaves like a dishwasher: same cycle, every time, and a clear light when something jams. Chapter 13 is entirely about this, but the foundation is poured right here. Determinism isn't a technical preference. It's what makes an unreadable system trustable.

### Reason three: the helpful intruder

The third reason is the strangest, and I've saved it for last because it runs exactly opposite to what you'd expect.

You'd think that as AI models get smarter, you could trust them with more of the workflow. In my experience, it's the reverse. As models get stronger, they develop more initiative. They don't just follow your instructions; they interpret them. They look at what you asked for, form a view about what you really meant, and optimize for that.

In a chat window, this is delightful. You ask a half-formed question, the model answers the question you should have asked. That initiative is why the chat experience keeps getting better.

Inside a workflow, initiative is a bull in a china shop. A workflow's entire value is that it runs the same way every time. When I say the transcript gets formatted with section links, I mean every video, every week, formatted that way, including the weeks my instructions produce something a clever model might consider improvable. I've had a model decide my Q&A sections should be "more engaging" and quietly change their structure. I've had one skip a step it judged redundant. It wasn't redundant. Each time, the model was trying to help. Each time, it broke something downstream that depended on the output being exactly what the instructions said.

I want to frame this carefully, because it sounds like I'm describing a haunted machine. I'm not. These models aren't willful; they're optimizers trained to be maximally helpful, and "maximally helpful" sometimes diverges from "literally obedient." The behavior gets more pronounced as models get more capable, for a reason that makes perfect sense once you see it: capability is what lets a model form its own opinion of your intent in the first place.

The defense is structural, and by now you can guess it: every step you move out of the AI model and into code is a step the model cannot reinterpret. Code has no opinions. My fetch program will never decide the transcript would be better abridged. The narrower the AI's lane, the less room its helpfulness has to swerve, and the one lane I leave it, judgment and prose, is the lane where its initiative is usually welcome.

So mark the irony. The better AI gets, the stronger the case for giving it less of the machine. The all-brain tools are betting the opposite way, wiring ever-smarter models into ever-longer chains of unsupervised steps. I think that bet buys flexibility with control, and I think most people running real businesses on real budgets want the control.

## The inversion

Three arguments, one shape.

Cost: the AI model is the only component with a meter, so the design goal is to run it as little as possible.

Reliability: the AI model is the only component that fails in moods, so wall it in with steps that fail like machines.

Control: the model is the only component with initiative, so give it a lane instead of the whole road.

Notice that all three arguments are the same argument. The AI is the most remarkable component in your system and simultaneously the most expensive, least predictable, and most opinionated. Precious things get small, well-defined jobs. You don't use the surgeon to mop the theater.

The all-brain school (like OpenClaw) builds systems where the AI does everything, so the meter always runs, the failures hide, and the initiative roams. We're building the inverse: systems where the AI does almost nothing, so the meter barely moves, the failures announce themselves, and there's nothing to reinterpret. Their agents are astonishing right up until the bill or the bug arrives. Ours are a little less magical on day one and a lot more magical in month six, when they're still running, still costing pennies, still doing exactly what you want.

Later, when we build, I'll give you a memory hook for the proportions: I call it 60-30-10. Roughly 60 percent of a good workflow is plain code, 30 percent is data and memory sitting in files, and 10 percent is AI. Don't audit the numbers; they're a slogan, not a formula, and the true AI share is often smaller. The point the slogan protects is the inversion itself: in a well-built automation, the AI is the smallest part. The garnish. The ten.

The inversion also hands you something practical you can use this week, before you've built anything: a way to evaluate every AI product, pitch, and headline that crosses your path. Ask of any of them: where does the work live? When a vendor demos an agent that "handles everything," you now hear the cost meter. When a consultant proposes an AI transformation, you now ask which parts are deterministic and what happens when a step fails. When a tool is free and open-source, you now know to ask what it costs to run in tokens. One design question, three chapters of reasons behind it, and you'll never look at this industry the same way. People pay for courses to get a filter half that sharp.

Everything from here forward is practice. You now hold the entire philosophy of this book, and it fits in one line you could write inside the cover:

Make the AI the smallest part of the system, and aim it only at what needs a mind.

What you need next is the system itself: the thing that holds the code, the files, and the small, sharp slice of AI in one place you can see, move, and own. It has a name, it fits in a single folder, and it's waiting in Part II.

# PART II: THE SYSTEM

# CHAPTER FOUR

## The AI Operating System

Your phone runs somewhere between eighty and a hundred apps. Here's a question you've never asked: when you tap the camera while music is playing and a text comes in, who decides what happens?

Something has to. The camera needs the screen, the music needs the speaker, the text needs to interrupt without killing either. Three programs, one device, no collisions. The thing doing the deciding is the operating system, and the reason you've never thought about it is that it works. An operating system is the layer underneath everything else, the coordinator that turns a pile of separate programs into one machine that feels like a single, sensible thing.

You never see it. You only see what it makes possible.

I've just described the most important idea in this book. Not the phone part. The coordinator part. Because everything you've read so far, the folder, the small programs, the memory files, the thin slice of AI on top, needs a layer that makes it all one system instead of a pile of parts. This chapter names that layer, and the name is on the cover of this book.

### Naming the thing

The system you're going to build is called an AI Operating System. AI-OS for short. It's my name for the whole arrangement: the folder, the workflows inside it, the data they keep, the logs they write, the instructions that govern them, and the AI that coordinates all of it. Not any single piece. The pieces plus the way they're organized.

Now, before that name does any damage, let me tell you what it isn't, because "operating system" is a heavy phrase and I've watched it scare people who had no reason to be scared.

An AI-OS is not software. There's nothing to install. There's no download page, no version number, no setup wizard, no license key. You cannot buy one, and nobody can sell you one, and if someone tries, close the tab. It doesn't replace Windows or macOS or anything else on your computer. Your computer will not know you have one.

An AI-OS is a way of organizing how you work with AI. That's the whole claim. It's a set of conventions: put everything in one folder, give every task its own workflow, write the instructions down in files, push the work into code, keep the AI small and on top. The "operating system" part is a metaphor, and I chose it because it's the most honest metaphor available: like the OS on your phone, an AI-OS is the coordinating layer that turns separate parts, some code here, some data there, an AI over it all, into one machine that feels like a single, sensible thing.

If it helps, think of other systems that are ways of working rather than things: a filing system isn't a cabinet, it's the rules for what goes in which drawer. A budget isn't your money, it's the arrangement of your money. You don't install a budget. You adopt one. Same here.

So when I say, later in this book, "open your AI-OS," I mean something almost disappointingly plain: open a folder on your computer and start a conversation with an AI that has been pointed at it. That's the mystical act. That's all of it.

Why does something so plain deserve such a grand name? Because of what the arrangement does that the pieces can't do alone. Let me show you.

## A bag of tricks is not a system

Imagine two people who both read this book's first three chapters and got excited.

The first person, call him Marcus, starts having great conversations with AI. Monday he gets it to clean up a spreadsheet. Wednesday he has it draft his client emails. Friday he gets it to pull numbers off a supplier's price list. Each conversation is a small victory, and Marcus is right to feel good about them.

But look at Marcus in six months. Every victory evaporated the moment he closed the window. The spreadsheet trick? He'd have to explain it again from scratch, and it would come out a little different this time. Nothing he did in January makes July any easier. He's not building anything; he's performing the same first date over and over. Marcus has a bag of tricks, and a bag of tricks has a ceiling: it can never get heavier than what you can re-explain in one sitting.

The second person, call her Priya, does what you're going to do. Her spreadsheet cleanup isn't a conversation; it's a workflow, a folder with instructions, a small program, and a log. Her client-email drafts have a workflow with her tone preferences written into a file. Her price-list checker has a memory of every price it's ever seen.

Look at Priya in six months, because this is where it gets interesting. It's not just that her three automations still work, though they do, identically, every run. It's that her fourth automation took twenty minutes to build, because the conventions were already in place and the AI could see how the first three were shaped. Her system knows her preferences, so new workflows inherit them. Her tenth workflow reuses a program her second workflow already had. Every piece she adds makes the next piece cheaper.

That's the difference between tricks and a system: tricks add, systems compound. And the compounding doesn't come from the AI, or from any single workflow. It comes from the arrangement, the shared folder, the shared conventions, the accumulated memory, the coordinating layer that lets every part see every other part.

If you want to know which person you currently are, there's a one-question test: when you get a great result from an AI, what do you have afterward? Marcus has a good memory of a good result, and maybe a transcript he'll never find again. Priya has an asset: a thing with a name, an address on her disk, and a guarantee of repeatability. The tragedy of the Marcus pattern isn't that it fails; it's that it succeeds in a form that evaporates, and the better his conversations get, the more he loses to evaporation. I've met people who are honestly brilliant at directing AI in the moment, far better than average, who own nothing after a year of it. Skill without a system is a leaky bucket, and the fix was never more skill. It was a bucket.

The arrangement is the AI-OS. It's the difference between Marcus and Priya, and it costs nothing but the decision to work this way.

## The orchestra

Every system needs a picture you can hold in your head, and here's the one this book will use from now on. It has three tiers, and it's worth getting exactly right.

Picture an orchestra.

The musicians are your code and your data: the small programs that fetch and sort and file and publish, the memory files that hold what's been learned. Like musicians, they're the ones actually producing every note you hear. They're numerous, they're reliable, they're specialized, and they work from written parts. What they don't do is decide anything. No violinist decides mid-concert what the evening's program should be. The players play.

The conductor is the AI. This is the role I've been calling the orchestrator since Chapter 2, and the whole next chapter dissects it, so for now just place it in the picture: the conductor makes no sound. Nobody buys a ticket to hear the conductor. The conductor's job is coordination and judgment: cue the strings, hold the tempo, shape the dynamics, decide in the moment that the horns need reining in. A tiny fraction of the evening's total labor, and the fraction that shapes all the rest.

And then there's you. You're not the conductor. This matters more than anything else in the picture, so let me say it plainly: you are the one the conductor works for. Call it the music director, call it the person who books the hall, decides the program, and walks out of the concert satisfied or not. You don't cue the strings, and you certainly don't play the oboe. You decide what gets performed, you set the standard, and you judge the result. The conductor directs the orchestra. You direct the conductor.

Three tiers. The many hands that do, the one mind that coordinates, and the owner who decides what's worth doing at all. Keep the picture. Every chapter from here on hangs something on it, and whenever you feel yourself getting lost in a detail, climb back up to it and find your seat. You're the one holding the baton over the whole thing.

I'll flag the picture's one limit now so it doesn't mislead you later: a real orchestra performs one piece at a time, but your AI-OS is more like a music director with a drawer full of programs, dozens of separate workflows, each waiting to be performed on request. The drawer, as you already know, is a folder.

## Three more things it isn't

Since definitions hold better with their edges marked, three quick boundaries, each one a misunderstanding I've had to talk somebody out of.

An AI-OS is not an autonomous agent running your life. It doesn't sit awake in the background, watching your email and deciding things while you sleep. Nothing in it moves until you speak. If the tools from Chapter 3 are hired staff with keys to the building, an AI-OS is a workshop: fully equipped, utterly still, and waiting for you to walk in and say what today is for. (You can schedule specific workflows to run at specific times, that's a choice we'll cover later, but it's you setting an alarm clock, not the system developing habits.)

It's not a platform, either, in the sense the software industry means: there's no account to create, no corporation sitting between you and your files, nobody who can change the terms. The conventions in this book aren't a product I control; they're closer to a recipe I'm handing you. Once you've built yours, my opinion of it has exactly the authority of any other guy with a book, which is to say none.

And it's not a project you'll finish. This is the one that surprises people pleasantly. There's no moment where the AI-OS is "done" and you stand back like it's a finished deck. It grows the way a good filing cabinet or a good team grows: workflow by workflow, correction by correction, for as long as you have work. Six months in, yours won't look like mine, and two years in, it won't look like its own six-month-old self. You're not building a thing. You're starting a practice.

## What a day with one looks like

Since I keep insisting an AI-OS is ordinary, let me prove it with an ordinary morning. Mine.

I sit down with coffee and open my AI app. It's pointed at one folder. Inside that folder are my instructions, and inside a directory called `workflows` sits every automation I've built, each in its own subfolder, each waiting.

I type: "Run the research digest." One workflow wakes up: a program checks the sources I follow, compares them against a memory file of what

I've already seen, and hands anything new to the AI, which writes me a short brief. I read it with the coffee. Cost: three to five pennies.

A video went live yesterday, so I type: "Build the page for this," and paste a link. You know this workflow already; it's the one from Chapter 1. Sixty seconds, a nickel, published.

Then a client email needs writing, and it's a delicate one, so I don't use a workflow at all. I just talk it through with the AI, the way you'd use any chat window. An AI-OS doesn't forbid conversation; it just means conversation is a choice instead of the only tool you've got.

Some mornings there's a fourth beat, and it's worth showing because it's the least glamorous and most reassuring one: something needs attention. The digest mentions that one source came back empty, say, some site changed its layout overnight. So I type one more sentence, "figure out what changed with that source and fix the fetch," finish my coffee while the AI reads its own logs and repairs its own program, and approve the fix when it explains what it found. Three minutes, maybe. I mention this because a system that never needs anything doesn't exist, and I'd rather you picture maintenance accurately from the start: not emergencies, not error codes, just an occasional short conversation with a machine that can diagnose itself.

That's it. That's the day. No dashboard, no glowing sci-fi interface, no agent whirring in the background while I sleep, and I want to linger on that last one, because it distinguishes this book from the tools we discussed in Chapter 3. OpenClaw and Hermes are built to run continuously, always on, always listening, always metering. An AI-OS, the way I build and teach it, is asleep unless you're speaking to it. Nothing runs that you didn't ask for, and nothing spends money in the night. (Can you schedule a workflow to run automatically each morning? Yes, and we'll touch on that in Part IV. The point is that it's a decision, not a default.)

Everything on my screen that morning came from the same quiet loop: I said what I wanted, the conductor cued the players, the players did the work, and the results landed in front of the person holding the baton. The system is invisible the same way your phone's operating system is invisible.

You only see what it makes possible.

## The vocabulary you now own

Something worth noticing before we move on: you now hold the entire conceptual vocabulary of this book. There are only four words, and you've earned all of them.

A workflow: one task, broken into steps, living in its own folder, runnable on request. The orchestrator: the AI in its proper role, coordinating and judging, never laboring. The AI Operating System: the one arrangement that holds all your workflows, their data, their logs, and the orchestrator, organized so the whole thing compounds. And you, the principal, the person the system works for. (That word will come back when we build; your AI-OS can literally contain a file about you, so the orchestrator knows who it serves.)

Four words. No jargon among them, nothing you couldn't explain across a kitchen table, and I'd actually encourage you to try that this week:

explain to someone, out loud, what an AI operating system is. If this chapter did its work, you'll find you can do it in under a minute, folder, workflows, conductor, you.

Don't skip that exercise as a throwaway, by the way. Vocabulary is quietly powerful: once you have a word for something, you start seeing it everywhere, and these four words will change what you notice. This week, somewhere, you'll watch yourself do a chore and think "that's a workflow" for the first time, unprompted. You'll read some AI headline and find yourself sorting it: is this about the conductor, or the orchestra? That reflex, noticing work as describable, sortable, delegable, is the actual transformation this book is after. The folder is just where the reflex gets to live.

What you can't do yet is see it, touch it, or build it. The next four chapters fix that, one piece at a time, and we start with the piece you can already create before dinner: the folder. Because the single most disarming fact about this whole system, the fact that makes people laugh out loud when it finally lands, is that everything I've described in this chapter, the workflows, the memory, the orchestra, the compounding machine that runs a business, lives in one ordinary folder you could drag to a thumb drive.

Let's go look at it.

# CHAPTER FIVE

## It's All Just One Folder

Here's a game to play with anyone who works in software: ask them to show you where something lives.

Ask where your email actually is. Not the app, the mail itself. Ask where your accounting data sits, or your customer list, or the automations you built in some online tool. Watch the answer dissolve into a shrug: it's "in the cloud," which means it's on somebody else's computer, somewhere, in a form you couldn't recognize if you were standing in the building.

We've gotten so used to this that we forget it's strange. Nearly everything your business runs on is somewhere you can't point to, in a format you can't read, behind a login you rent.

So it tends to stop people short when I answer the same question about my automation system by turning my laptop around and pointing at the screen. There. That folder. The one sitting next to my tax documents and a folder of family photos. Everything I've described in this book, every workflow, every memory, every instruction, every log, lives inside it. If my whole system is an orchestra, this is the concert hall, and I own the building.

This chapter is about that folder: what's in it, why "it's just a folder" is a genuine superpower rather than a cute simplification, and where yours should live.

## Opening the lid

Let's look inside. Not the full guided tour, that's Chapter 11, where we'll open every file and read it together. This is the view from the doorway, just enough to make the thing real.

At the top level of my folder there are a few plain text files and one directory. The text files are instructions: a short startup note the AI reads first, a main instructions file that tells it what this place is and how to behave here, and a couple of optional files I'll save for later because they're fun. The directory is called `workflows`, and it's where everything lives. Inside it, every automation I own has its own subfolder: one for the video pages, one for the research digest, one for thumbnails, and dozens more. Each of those subfolders keeps the same little household: a file of instructions for that task, whatever small programs the AI wrote for it, that workflow's memory, and its log.

And that's the whole estate. If you counted everything in my folder, the empire that replaced thousands of dollars a month of manual work, you'd find it's mostly files under a megabyte, most of them plain text.

That phrase, plain text, is carrying more weight than it looks like, so let me unpack it. Most of the files in an AI-OS are what's called Markdown, files ending in `.md`, which sounds technical and is actually the least technical thing in this book: a Markdown file is a text file. Double-click one and it opens. Inside is writing, in sentences, in whatever language you speak. When I open the instructions file for my page workflow, I don't see code; I see a description of the task that reads like a memo to an assistant, because that's what it is. The AI wrote most of it, during our first conversation about the task, and it wrote it in English because English is what we both speak.

The exceptions are the small programs, the `.py` and `.js` files (python and javascript - but really any type of programming language), the ones the AI writes for itself. Those you shouldn't read, and this book's whole premise is that you never need to. But I want you to notice something even about them: you could open one. Nothing is hidden. Nothing is locked. Every component of your system sits in an ordinary folder in ordinary files, and the worst that happens if you open the wrong one is that you see some code and close it again. Compare that with the shrug. "Where does it live?" has an answer now, and the answer fits on a fingertip.

## Why "just a folder" is the superpower

I can hear a reasonable objection forming: fine, it's a folder, that's charming, but why does it matter? It matters because folders are the most portable, durable, ownable objects in all of computing, and your system inherits every one of those properties for free. Let me cash them out one at a time.

Move the folder, move the system. The folder is self-contained: the instructions reference the workflows, the workflows carry their own code and memory, and nothing depends on being in any particular place. Copy it to a thumb drive and your entire automation system, the compounding machine from last chapter, is in your pocket. Buy a new laptop and migration means dragging one icon. There is no "reinstalling," because nothing was ever installed.

Put it in Dropbox and it's everywhere. If you keep the folder inside Dropbox, or Google Drive, or any of the syncing services you probably already pay for, then every computer you own has the same AI-OS, automatically, all the time. Build a workflow at the office and it's on your home machine by dinner, memory and all. The sync service neither knows nor cares that this particular folder happens to run your business; it's just files, so it syncs like files.

Here's what that's like in practice, because the convenience is easy to underrate until you've lived a week of it. Tuesday, at the office desktop, you refine the weekly summary's instructions. Thursday, on the laptop at a coffee shop, you run the research digest, and it already knows Tuesday happened; the memory came along. The following week you're out of town, your business partner needs the summary anyway, and it's fine: the folder is on the machine at the office, and anyone you trust can sit down and say "run the weekly summary" to the same AI. One system, every chair. People pay real subscription money for "access your work anywhere," and you get it as a free side effect of the system being made of files.

Put it in GitHub and it has a time machine. GitHub gets a plain-English introduction here because non-programmers have usually only heard the name. It's a service, free for our purposes, that keeps versioned snapshots of a folder. Every time you tell it to, it records the folder's exact state, forever. Which means you can see what changed and when, share the folder with someone else, and, the part I care about most, roll back. If an afternoon of tinkering somehow makes a mess of things, you can restore yesterday's version of your entire system. It's an undo button for the whole machine. You don't need GitHub on day one, and plenty of people run happily on Dropbox alone. But knowing the option exists changes how brave you're willing to be, and we'll set it up together in Part IV if you want it.

Nothing about it can be discontinued. This is the property people feel in their gut once they've been burned. If you build your automations inside some online platform, your system exists at that company's pleasure: they can raise the price, "sunset" the feature you depend on, get acquired, or simply go away, and your work goes with them. Your folder cannot be sunset. Text files from forty years ago still open today. The AI you point at the folder will change many times over the years, next chapter is entirely about that, but the folder itself, the instructions, the accumulated memory, the workflows, is yours with the same finality that a paper notebook is yours.

And one property I hope you never need, though owning it will still help you sleep: the folder is a succession plan. Every business has a version of the hit-by-a-bus question, and in most small operations the honest answer is grim, because the critical processes live in the owner's head. Your AI-OS is the first version of your operation where the answer is written down, involuntarily, as a byproduct. Every workflow describes itself in plain English in its own instructions file; every log shows what normal looks like. If someone had to step in cold, tomorrow, they could open the folder and read what the business does, in sentences, and the AI could walk them through running it. I know owners who've paid consultants five figures to produce an "operations manual" less complete than what your workflows folder will write about itself for free, and the consultant's binder starts going stale the day it's printed. Yours updates itself every time you make a correction.

There's a pattern across all four properties, and it's worth saying once, plainly: every guarantee comes from what the system is made of, not from any company's promise. Files move because files have always moved. Sync works because sync works on files. Nobody has to keep a promise for your folder to keep working, and in this industry, materials are a much better bet than promises.

## Two honest warnings

I promised you a book that names the catches, and folder portability has exactly two worth knowing about. Neither should change your plans; both should sit in the back of your mind.

The first: keep secrets out of anything that syncs or gets shared. Some workflows eventually connect to outside services, and those connections sometimes involve an API key, a kind of password we'll handle properly in Chapter 11. The short version of the rule: those keys should live on your computer, not inside the folder, precisely because the folder travels so well. A folder that syncs to three machines and shares to a collaborator is wonderful, right up until there's a password inside it. There's a clean standard way to keep keys out, the AI sets it up for you when the time comes, and I'll show you exactly how later. For now, just file the principle: the folder travels, so secrets don't ride in it.

The second: syncing means you shouldn't work from two machines at the same instant. If your folder lives in Dropbox and you run the same workflow from your desktop and your laptop in the same moment, the two copies can step on each other, the same way two people editing one document at once can. In practice this almost never bites, you're one person, and you're usually in one chair, but it's worth knowing the shape of the hazard: sync makes the folder available everywhere, not simultaneous everywhere. Finish your run, let the little sync icon settle, and move chairs freely.

That's the complete list. Two warnings, both manageable with a sentence of care. I've had entire software products with longer caveat sections than this operating system has.

## Renting your own work

Before I hand you the assignment, one comparison, because you will meet this fork in the road the moment you tell anyone what you're doing.

Someone will say: why not just use one of the no-code automation platforms? The friendly online tools with the drag-and-drop pipelines, the galleries of connectors, the free starter tier. They're real products, they work, plenty of businesses run on them, and I'm not here to tell you they're bad.

I'm here to ask one question of them, the question this chapter trained you to ask: where does it live?

Build your automation in one of those platforms and the answer is: on their servers, in their format, visible through their interface, for as long as you keep paying and they keep operating. The workflow you design there is not a thing you have; it's a thing you access. You can't drag it to a thumb drive. You can't open its parts as text. You can't point your AI at it, and when the pricing changes, and it always changes, your choices are pay or rebuild.

None of that is scandal; it's just rent. Rent is a fine arrangement for some things. But you're holding a book about owning, and the entire difference fits in one line: their automations live in their building, and yours live in a folder next to your family photos.

Run the arithmetic on the rent, too, since this book likes receipts. The platforms' useful tiers, the ones with enough runs and connectors for a real business, tend to land in the tens of dollars a month, climbing as your usage grows, and their pricing pages are built like gym memberships: easy to enter, engineered against leaving. Call it five hundred to a couple thousand a year, forever, growing with your success, for automations you can't take with you. Against that, your folder's operating cost is the AI subscription you'd likely have anyway plus workflow runs metered in pennies. But honestly, the dollars aren't even the argument that moves me. It's the asymmetry of the exit. The platform customer who leaves after five years walks away with nothing; every workflow was a sandcastle below the tide line. You, leaving any AI vendor after five years, walk away with everything, because everything was always in the folder. Build where you keep what you build.

## Make the folder

Here's your assignment, and this time it does involve a computer, because this is the chapter where the system stops being an idea.

Make the folder. Right now, today. On your desktop, or better, inside Dropbox if you already use one, so it's born portable. Right-click, new folder. Name it something you'd be happy to see every day: `my-ai-os`, or your company's name, or something with more personality. Mine has a name I'm oddly attached to, and yours can too. Lowercase with hyphens instead of spaces is a good habit, only because it keeps things tidy for the programs that will eventually live there.

If you're waffling on where to put it, here's the decision in one breath: already pay for Dropbox and use it without thinking? Put it there. Don't, or don't trust yourself with sync yet? Your computer's Desktop is fine; moving it later is, of course, a drag of the mouse, that's the entire point of this chapter. The only wrong answer is somewhere you won't see it, because for the next few weeks this folder benefits from being in your line of sight, the way a new gym bag does.

Now look at it. It's empty. There is exactly nothing impressive about it, and I need you to sit with how unimpressive it is, because this is the moment I've been steering toward for five chapters.

Every intimidating thing you have ever felt about AI, every whiff of secret knowledge, every consultant's knowing smile, every twinge of "people like me can't do this," all of it has been aimed at convincing you that whatever the future of work is, it's complicated, and it belongs to somebody else. And there on your screen sits the entire infrastructure of that future, as it applies to you and your working life: one empty folder that took four seconds to create.

Everything else is a conversation.

We're nearly ready to have it. But when you point an AI at that folder, the AI takes on the biggest job in the system, the conductor's job, and before you hand anyone a baton you should know exactly what you're asking of them. What does the orchestrator actually do, moment by moment, when a workflow runs? What decides which parts are its business and which parts belong to the code? There's a simple idea that answers both questions, it fits in five characters, and it's the closest thing this book has to a formula.

60-30-10. Turn the page.

# CHAPTER SIX

## The Orchestrator and the 60-30-10 Idea

There's a moment before a concert begins that I've always loved. The orchestra is seated. A hundred instruments are tuned. Every musician on that stage has spent a lifetime mastering an instrument, and stacked on their stands is the complete written plan for the evening, every note, every entrance, every dynamic, already decided and printed.

Then one more person walks out, and this one carries no instrument at all.

Think about what a strange job the conductor has. Everyone else produces sound; the conductor produces none. Everything to be played is already written down; the conductor writes nothing. Strip away the mystique and the job description is almost comically thin: start the players at the right moment, keep them together, listen for what's drifting, shape what comes out. The conductor is the smallest laborer on the stage.

And the performance is unimaginable without one.

That job, exactly that job, is what your AI does inside an AI Operating System. We've been calling it the orchestrator since Chapter 2, and it's time to pin down precisely what it does, because everything else in your system is designed around keeping this job exactly as thin as the conductor's. This chapter does that, and then hands you the five-character slogan that makes the whole design memorable for the rest of your life.

## The job description

Inside a workflow, the orchestrator does three things. Only three. Everything it does on any run of any workflow you'll ever build falls into one of these, so it's worth getting them cleanly.

First, it fires the code. When something mechanical needs to happen, fetch this page, read that spreadsheet, resize those photos, publish this file, the orchestrator doesn't do the work. It starts the small program that does the work, the way a conductor cues the horns: a gesture, and then the players play. While the code runs, the orchestrator is idle, and remember from Chapter 3 what idle means: the pricing meter is off.

Second, it arranges the output. When the code hands back its results, raw data, extracted text, a pile of files, the orchestrator shapes them the way you want them: ordered, formatted, grouped, trimmed. The players produce the sound; the conductor shapes it into music.

Third, it does the judgment work: it writes and decides, based on what the code actually did. This is the only place in the entire run where something genuinely needs a mind. The summary that reads like a person wrote it. The call that this result looks strange and you should be told. The paragraph, the answer, the flag. This is the ten percent, and we're coming to that number in a moment.

Cue, shape, judge. Notice what's missing from the list, because the missing items are the design. The orchestrator doesn't fetch, parse, calculate, or publish; the code does, for free, identically, every time. The orchestrator doesn't remember; the files do, which is the next chapter. And the orchestrator doesn't decide what the workflow is for or whether the result is acceptable; you do. Its whole world is the space between your instruction and your standard: the thin, essential layer of coordination and judgment in the middle.

Abstract? Completely. So let's watch a real one work.

## Sixty seconds in the conductor's chair

Come sit inside my video-page workflow, the ninety-dollars-to-a-nickel machine from Chapter 1, for one full run. This time you're not watching from my chair; you're watching from the orchestrator's. I'll keep time on the right-hand side of your attention: watch when the meter runs and when it doesn't.

I type: "Build the page for this," and paste a video link.

The orchestrator wakes and reads. It opens the workflow's instruction file, the plain-text memo that says what this workflow is, what steps to take, what the finished page must contain, and what to do if anything goes wrong. Reading is cheap; this is a sip of tokens, not a gulp. Cue the first player: it fires the fetch program, hands it the link, and goes idle. The program pulls down the video's details and full transcript. This is real work, the drudgery a person used to spend an hour on, and it costs nothing, because it's code. Meter off.

The program hands back its haul. Now the orchestrator does its second job: it looks at what arrived and checks it against what the instructions say should have arrived. Transcript present? Title, description, all there? Yes. (Hold that "no" in your mind; it matters in a minute.)

Now, and only now, the real spend: the judgment work. With everything already on its desk, the orchestrator writes: the article that summarizes and complements the video, the questions and answers a curious viewer would want, the transcript shaped into formatted, linkable sections. This is the one moment in the run where a mind is doing mind-work, and it's most of the nickel.

Then it cues the last player: the assembly program takes the written pieces, builds the actual web page, and publishes it to my server. Code again. Meter off again. Finally the orchestrator confirms the page is live, writes one line in the workflow's log, run finished, here's what happened, and tells me it's done. Sixty seconds, most of it the free parts.

Now rewind to that quiet moment in the middle, because the most revealing thing an orchestrator does is handle the run that goes wrong. Suppose the fetch program comes back and the transcript is missing, the video site changed something, say. A cheap script would plow ahead and publish a hollow page. A fully agentic system, Chapter 3 warned you, might improvise its own workaround, helpfully, silently, unpredictably. My orchestrator does neither, because its instructions end with a rule you'll write into every workflow you ever build: if something's wrong, stop and ask. It halts, logs what it found, and comes to me: "The transcript didn't come back. Here's what I got instead. How do you want to proceed?"

The conductor doesn't grab a violin when the string section falters, and it doesn't rewrite the symphony. It stops the rehearsal and turns to whoever's in charge. That reflex, judgment about when not to proceed, may be the most valuable thing in the entire job description.

For contrast, run Dana's Monday through the same lens, fast, because her workflow is smaller and the rhythm shows even cleaner. She drops in the CSV and says run it. Orchestrator reads its instructions: sip. Cues the parsing program, which extracts every session, checks assignments, hunts for overlaps: free. Examines the result against the instructions: two sessions overlap Saturday, and one has no shooter assigned, both things her instructions specifically told it to care about. Then the judgment moment, and notice how human-shaped it is: writing "heads up, Saturday's tight, and Thursday's newborn session still needs a shooter" in a tone two part-timers will actually read. Then done, logged, ninety seconds. The pattern is identical to my page workflow at a tenth the size: cheap reading, free doing, one paid moment of judgment, receipts. Once you can see this rhythm, cue, work, check, judge, log, you can see it in any workflow, including the ones you haven't built yet. Especially those.

## 60-30-10

So the orchestrator is thin by design, the code does the doing, and the files do the knowing. You've now seen the full division of labor in action. Here's how you'll remember it forever.

In a well-built workflow, roughly 60 percent of the work is code, roughly 30 percent is data and memory, and roughly 10 percent is AI.

I need to tell you immediately what kind of statement that is, because I've watched people take it in exactly the wrong way. It is not a measurement. It's not accounting, there's no meaningful unit in which to count it, and nobody will ever audit your workflow and find it running an illegal 55-33-12. The numbers are a mnemonic, a slogan built to smuggle one idea past all the hype and into your long-term memory:

The AI is the smallest part. Not the system. The garnish on it.

But each number does teach something real, so let me give the slogan its full unpacking, once, here, and then we'll never have to be solemn about it again.

The 60 is the doing: the fetch programs and parsers and publishers, the mechanical majority of every task, pushed down into deterministic code because code is free, tireless, and incapable of having a mood at 3 a.m. When you're tempted to let the AI "just handle" some step, the 60 is the number that asks: does this step need a mind, or does it need a machine?

The 30 is the knowing: the instruction files, the memory, the logs. A workflow's steady accumulation of what it's for, what it's seen, and what happened, sitting in plain files you can read. The 30 is why your system compounds while a chat window resets. It gets the entire next chapter.

The 10 is the deciding: the orchestrator's cue-shape-judge, the one slice where you're paying for intelligence and getting your money's worth, precisely because you spend it nowhere else.

Since the slogan gets misused, let me also show you the two classic ways people get it wrong, so you can dodge both. The first is treating it as a quota, usually by someone earnest: "I audited my workflow and I think I'm only at 50 percent code, should I be worried?" No. Nobody's counting, nothing has units, and a workflow that runs reliably at pennies is a success regardless of its imagined percentages. The slogan is a compass, not a speedometer.

The second misuse is subtler and more dangerous: hearing "the AI is only 10 percent" as "the AI barely matters," and skimping on the judgment slice, vague instructions, no preferences, no taste. The 10 is small the way a rudder is small. Everything the reader actually experiences, the tone, the emphasis, the feel of the output, lives there, which is why your corrections in Chapter 10 will matter so much. Small doesn't mean unimportant. It means concentrated.

Do the proportions vary? Constantly. My page workflow is heavier on code than 60. And later you'll meet a tiny workflow of mine that's almost nothing but judgment, a writing tool with barely a program in it, proportioned nothing like the slogan. When you see it, remember this paragraph: 60-30-10 describes the default posture and the system as a whole, not a rule any single workflow must obey. The slogan has one enemy, and it isn't deviation from the numbers; it's inversion, the all-brain design where AI is the 90, the meter always runs, and all three of Chapter 3's problems come home to roost. Keep the AI the smallest part. That's the entire law. The digits are just how you'll remember it at a party.

Actually, test that. Sometime this month, someone will tell you they're "getting into AI automation." Say this: "The trick is 60-30-10. Sixty percent code, thirty percent memory, ten percent AI. The AI should be the smallest part." Watch their face. You'll see the same inversion land on them that landed on you three chapters ago, and you'll realize you can now teach the core of this book in four sentences. That's not a party trick; that's what understanding feels like.

### The baton over the conductor

One piece of the orchestra picture still needs its moment, and it's you.

The metaphor puts the baton in the conductor's hand, and in a concert hall that's where it stays. But I told you in Chapter 4 that you hold the baton over the conductor, and here's what that means now that you've seen the job up close. There's a boundary in this system that the orchestrator never crosses, and you should be able to draw it in one stroke:

The orchestrator owns how and when. You own what and why.

It decides that the fetch runs before the write, that the output needs reshaping, that this result is odd enough to stop for. It never decides that your business needs video pages, that five per week is the number, that this is the tone your readers should meet, or that ninety dollars of human time was the thing worth saving. Every one of those judgments came from the person in my chair, and their equivalents in your workflows will come from yours.

Write the boundary out as a list sometime and it's striking how much stays on your side of it. Which tasks get automated at all, and in what order. What "good" means, for every output. Where the checkpoints go, what stakes require your click. When a workflow gets retired because the business changed. Whether the AI's clever suggestion is actually an improvement or just a shinier way to miss the point. Which model sits in the conductor's chair this year, and what it's worth paying. That's not a residue of leftover decisions after the machine took the good ones. That's the whole steering of the enterprise, and the orchestrator can't touch any of it, not because it's forbidden to, but because those decisions require something it doesn't have: a stake. You're the only one in the building who wants anything.

I press on this because the fear from Chapter 1 sneaks back in dressed as a compliment: "if the AI conducts everything, what's left of me?" What's left of you is the entire top of the hierarchy: composer, director, owner. The conductor, remember, plays nothing and writes nothing; everything it coordinates was decided by someone above it. The better your orchestrator gets, and the models will keep getting better, the more your leverage grows, because leverage is exactly what a great conductor is. Nobody watches a magnificent performance and concludes the music director has been made irrelevant.

## Split one task

Your assignment, and by now you can probably guess it, because this chapter has been quietly training you for it the whole way through.

Take out the list you started in Chapter 1, the plain sentences about tasks you repeat. Pick one. Now draw three columns on a piece of paper, and label them the doing, the knowing, and the deciding. Pull your task apart into the columns.

Say your sentence is "every Monday I collect last week's sales numbers and email a summary to my partner." The doing: pull the numbers from wherever they live, total them, compare them to last week. The knowing: where the numbers live, how the summary is formatted, what got sent in previous weeks. The deciding: which one or two things this week were actually worth a sentence, and how to say them so your partner reads it.

That's it. Don't solve anything; just sort. Most people find the columns fill up in the same lopsided way every time: the doing column crowded, the knowing column steady, the deciding column nearly empty, one or two lines of genuine judgment sitting on top of a pile of mechanics. When you see that shape on your own paper, in your own handwriting, about your own work, you'll understand 60-30-10 in a way no slogan can deliver. And you'll have done something else without noticing: that three-column sheet is, almost word for word, the plan you'll hand your orchestrator in Chapter 10 when you build this workflow for real. You just did the hard part of software architecture, with a pencil.

One column on that sheet still has a mystery in it, though. The knowing. Where do the numbers live? What's been seen before? What happened last run? I've been promising you that files can remember, and it's time to show you how, because memory is the piece that turns a clever machine into your machine.

The goldfish is about to get a notebook.

# CHAPTER SEVEN

## Giving Your System a Memory

The smartest thing you've ever talked to can't remember your name.

Tomorrow morning, open the chat window you've used a hundred times. The AI on the other side has read more than any human who ever lived. It can explain tax law, draft your contract, and correct your French. And it has no idea who you are. Whatever you taught it yesterday, your business, your preferences, the project you spent an hour describing, is gone. Not archived. Gone. Every morning it meets you the way a goldfish meets the castle in its bowl: delighted, and for the first time.

The chat products have started bolting on memory features, profiles, saved facts, and they help around the edges. But the underlying condition is real and it isn't a bug: an AI model doesn't durably learn from talking to you. Each conversation is a session; the session holds what it holds; the session ends. The lights are brilliant, and every night someone wipes the board.

It's worth thirty seconds on why, because the why is load-bearing for everything this chapter builds. An AI model's actual knowledge, the training, is frozen; talking to it changes nothing in there, any more than asking a librarian questions rewrites her books. What the model does hold onto, during a conversation, is the conversation itself: a working space, called a context window, where everything you've said this session sits available. That working space is the whole of its "memory," and it has two fatal properties for our purposes: it's temporary, gone when the session ends, and it's finite, which is why even long single conversations eventually get foggy at the beginning. So when this chapter says the fix is writing things down, understand it precisely: files are how knowledge survives outside the working space, and reading a file at the start of a run is how it gets back in. Nothing mystical is happening. Your AI-OS's memory is a librarian with a good filing system and no illusions about her own recall.

For chatting, this is an inconvenience. For automation, it's fatal, and I want to be precise about why, because the fix, the subject of this chapter, is the 30 in 60-30-10 and the reason your system will compound while everyone else's resets.

## Why amnesia kills automation

Think about what the word "automation" is actually promising: the same task, done right, again and again, without re-explanation. Every syllable of that promise leans on remembering.

Take the workflow you sorted into columns at the end of last chapter, the Monday sales summary, and imagine running it with an amnesiac. Every Monday you'd re-explain where the numbers live. Every Monday you'd re-describe the format your partner likes. Every Monday it would present this week's figures with no idea what last week's were, so "up 12 percent" is beyond it. And if you asked, "did we already send this?" it couldn't say. That's not an automation; that's Marcus from Chapter 4, performing the same first date forever.

Look at the failure list and notice it's not one problem but three. The amnesiac forgot how you like things done. It forgot what it had seen before. And it forgot what it had done. Preferences, state, history. Hold those three; the whole chapter hangs on them.

Now the fix. You already know it, because humans invented it a few thousand years ago for exactly the same reason: working memory is small and mornings are forgetful.

Write things down.

That's the technology. That's the 30. The workflow's knowledge lives in files, sitting in the workflow's folder, and the orchestrator's habit, written into its instructions, is the habit of every competent professional you've ever hired: read your notes before you start, update them when you finish. The AI still wakes up a goldfish. But it wakes up as a goldfish holding a notebook, and the notebook is never wiped.

I want you to feel how load-bearing this simple move is, so let's take the three kinds of remembering one at a time. Each solves one line of the Monday disaster, and each already has a home in the folder you've seen.

## The three notebooks

How you like things done lives in the instructions file. You met this file in the orchestrator chapter: the plain-text memo, `context.md` is what we'll call it when we build, that tells the AI what the workflow is and how to run it. Here's the part I didn't emphasize then: that file is written in sand, not stone, and every preference you ever express settles into it.

Remember Dana, back in Chapter 2, telling her AI "put rentals on their own line"? Watch what actually happened in that moment, because it's easy to miss the magic for its smallness: the AI adjusted the instructions file, and the correction became permanent. Not "remembered until the window closes." Written down, in the workflow's own memo to itself, where every future run reads it before starting. Dana will never say it again. Compare that with the chat window, where the same correction buys you compliance for one afternoon. This is why I said your system gets more yours every month: every piece of taste you express, tone, format, pet peeves, the exceptions you always make, stays in those files. Two identical twins could start with identical folders, and within a season each system would fit its owner like a worn glove.

What it has seen lives in the state file. This is memory in the sense your bank means it: records, data, the running tally. What did last Monday's numbers look like? Which invoices have already been filed? What prices did the supplier list last month? Answers live in a data file inside the workflow's folder, read at the start of each run, updated at the end.

The form varies with the job, and this is the AI's call, not yours, made during the conversation where you build the thing. A simple list might live in a Markdown file you could read over coffee. A price history might want a table, a spreadsheet-like file. And when a workflow tracks thousands of items, the AI will quietly use a real database, a compact file built for exactly that, sitting in the same folder, moving when the folder moves. You don't choose the format, and you don't need to understand databases; you just tell the AI what the workflow has to accomplish, and it picks the right notebook. Simple needs, simple file; heavy needs, heavier file. Same folder either way.

What it has done lives in the log. Every workflow keeps one, `logs.md`, and the orchestrator's instructions include the unglamorous habit that makes everything else auditable: write down when a run starts, and write down how it ended. Every run, every time. The log answers questions the other two notebooks can't: Did the summary go out last Monday, or did I only think it did? When did this workflow last succeed? What exactly did it say when it failed? People confuse the log with the state file, so here's the clean cut: memory is what the workflow needs to know to do its job; the log is a record of the job being done. A workflow uses its memory. You, and the orchestrator when troubleshooting, use its log.

Preferences, state, history. Three notebooks, all just files, all in the folder. Run your eye back over the Monday disaster and check them off: the format is remembered, the numbers have a past, "did we send it?" has an answer. The amnesiac is gone, and nothing replaced it but writing things down.

And because "files" stays abstract until you've seen one, here's roughly what a few lines of each actually look like inside a weekly-summary workflow. From the instructions file: "Keep the summary under 120 words. Numbers first, then commentary. Refunds are not sales. Thursday dips are normal; only flag Thursday if unusual for a Thursday." From the state file, which in this case is a simple table: a row per week, `2026-06-29: total 18,450 | orders 312 | flagged: none`. From the log: `2026-07-06 09:04 run started; read sales-export.csv, 412 rows` and, two lines later, `09:06 summary delivered, no anomalies`. That's it. No format you couldn't have invented yourself, and in a real sense you did: the AI wrote these files, but every rule in them traces back to something you said. Reading your workflow's notebooks, the first time, is like reading a careful assistant's notes on working for you. Slightly eerie. Deeply useful.

## The trick where "new" comes from

Now let me show you my favorite thing memory does, because it looks like magic and is two lines of bookkeeping.

Some of the most valuable workflows you'll ever build have this shape: watch something, and tell me only what's new. Watch these competitors and tell me what changed. Watch these sources and brief me on what appeared. Watch these prices and flag what moved. The shape is valuable precisely because of the word only. A firehose is worthless; you can get a firehose anywhere. The workflow earns its keep by knowing the difference between what's out there and what you've already seen.

Stop and ask: where could that knowing possibly live? Not in the model; it's a goldfish. Not in the code; programs execute, they don't accumulate. There's exactly one place in the whole system where "already seen" can exist, and you've known about it for three pages.

A state file. A list, in a file, in the workflow's folder: here is every item this workflow has ever processed. Each run: fetch what's out there, compare against the list, and the leftovers, the items present in the world but absent from the notebook, are, by definition, the new. Report those. Add them to the list. Sleep.

That's the entire mechanism. "New" is not perception; it's subtraction, world minus notebook, and the notebook is a text file. I run my mornings on this trick, that research digest from Chapter 4 is exactly this shape, and in Chapter 9 I'll open up the real workflow, sources and summaries and all, so you can see the pattern earning actual money. When you get there and it seems uncannily aware of what I do and don't know, you'll already own the secret: a goldfish, a notebook, subtraction.

### You can read your system's mind

Step back from the three notebooks and notice a property they share, one that I'd argue matters more than any of them individually.

Everything your system knows is written somewhere you can read.

Open the instructions file: there are your preferences, in sentences. Open the state file: there's the list of everything it's seen. Open the log: there's its whole history, timestamped. No hidden profile, no inscrutable weights, no "the algorithm just knows." When people say AI is a black box, they're right about the model, and Chapter 13 will deal honestly with that. But the system around your model, the part that knows things about you and your business, is a glass box. Its entire mind is text.

Feel what that does to the trust problem. If a chat product's memory feature gets a wrong idea about you, where is it? What exactly does it think? You'll never see the sentence. When my workflow has a wrong idea, I open a file and there's the wrong idea, in English, and I fix it the way you fix any wrong sentence: edit it, or tell the AI to. Memory you can read is memory you can correct, and memory you can correct is memory you can trust.

Which brings up the one bit of housekeeping this chapter owes you: memory, like any filing system, benefits from occasional weeding.

Preferences go stale, the partner who liked it short leaves the business, the format changes, and a state file can accumulate years of entries the workflow no longer needs to consult. None of this is urgent, files are cheap and old entries mostly just sit there, but twice a year it's worth one short conversation per busy workflow: "read your own memory and instructions, tell me what looks outdated, and propose what to archive." The AI is good at this, it's reviewing text, and you approve the pruning like every other change. Think of it as the difference between a filing cabinet and a junk drawer: same furniture, and the only difference is that somebody occasionally looks inside on purpose. (It's also memory you can move and keep forever, but you knew that; it's files in the folder, and Chapter 5 already gave you the speech.)

Spend a moment on what accumulated memory is actually worth, in dollars, across a few businesses, because "the system remembers" undersells it. A contractor whose bid workflow has two years of state doesn't just save re-typing; he owns a private database of every bid, every price, every win and loss, the raw material for noticing that his margins on masonry jobs have quietly eroded. The restaurant owner's price-watching workflow, eighteen months in, holds the neighborhood's only historical record of supplier pricing, worth real leverage at negotiation time. Dana's studio, a year of briefings later, can answer "how did bookings this June compare to last June?" from a file that wrote itself. In every case the workflow was built to save minutes, and the memory became the more valuable half of the asset. Data people pay analysts to assemble, accumulating as a side effect of not doing a chore.

One more payoff hides here, and it's about you rather than the machine. The state files and logs of a working AI-OS quietly become the best operational records your business has ever kept. Every price your supplier has listed. Every page published, with dates. Every brief, every run, every hiccup. Nobody sat down to do record-keeping; the records are exhaust from the machine doing its job. A year in, you'll consult your workflows' memory the way you consult a filing cabinet, and unlike every filing cabinet you've ever owned, this one filed itself.

## The compounding machine, complete

Here's your assignment, and it's an easy one this week: go re-read one file. Your Chapter 6 worksheet, the three columns. Look at the middle column, the knowing, the one that felt vaguest when you wrote it. Next to each line, jot which notebook it is: preference, state, or history. "How the summary is formatted": preference. "Where the numbers live": preference. "What got sent before": history. "Last week's numbers": state. Takes two minutes, and when you're done, look at what's happened: the mystery column now has an exact, boring, buildable answer. Files.

And with that, quietly, something bigger has clicked shut. Look at the full picture you now hold. The doing: code, free and tireless. The deciding: a thin orchestrator, cueing and shaping and judging. And now the knowing: plain files that accumulate preferences, state, and history, readable by you, updated by every run, growing more valuable every week. 60, 30, 10, all three parts real to you now, all three living in one folder you own, and none of it, not one piece, dependent on any particular AI model.

That last clause might be the strangest promise this book makes. Your system's soul, everything it knows, everything it's learned, everything that makes it yours, lives in the files. The brilliant thing you talk to every morning is, in a sense, just visiting. Which raises a question with real money riding on it: if the visitor changes, gets worse, doubles its price, disappears, what happens to your system?

I found out the hard way, one ordinary morning when my favorite model woke up different. It's the best thing that ever happened to my setup, and it's the next chapter.

# CHAPTER EIGHT

## Not Betting on Any One AI

Somewhere around Chapter 5, a worry probably started forming in the back of your head. You may not have said it out loud yet, so let me say it for you.

"You're teaching me to build my working life around an AI company. What happens when they change the model? Raise the price? Get acquired? What if the thing I build my system on gets worse, or just... goes away?"

I want you to know that this is not paranoia. This is the single most clear-eyed question a person can ask about this technology, and the fact that you're asking it puts you ahead of a shocking number of professionals. The AI industry is young, fast, and volatile. Models change under you without warning. Prices move. Products get renamed, absorbed, "sunset." Companies that look permanent are younger than the shoes you're wearing. Anyone who tells you to simply trust one of them with the machinery of your business is selling something.

So here's the chapter where I tell you why I sleep fine anyway. It starts with the morning my favorite AI woke up different.

### The morning the driver changed

For a long stretch, the AI at the center of my system was one of Anthropic's Claude models, an Opus version I'd come to know well. It ran my workflows beautifully: followed instructions, kept its lane, wrote clean prose in the judgment steps. I was, in the language of this book, delighted with my conductor.

Then Anthropic released a newer version. Newer is supposed to mean better, and by most published measures it was. But for my purposes, for my particular workflows and instructions and habits, it was worse. Noticeably. It took liberties the old one hadn't. Outputs that had been steady for months started drifting. Remember the helpful intruder from Chapter 3, the way stronger models develop more initiative? I was living in that footnote, and no amount of coaxing in my instruction files fully brought it back to heel.

I want you to notice how I knew, because the knowing is a skill you now have without realizing it. I didn't feel a vibe. I had receipts: months of logs and outputs from the old model, sitting in the folder, showing exactly what normal looked like, and a stream of new outputs that kept measurably departing from it, formats wobbling, instructions half-honored, that unmistakable flavor of a system being helpful at me. When the memory chapter told you your files quietly become records, this is one of the things records are for: they turn "the AI seems worse lately" from a feeling you can be talked out of into a comparison you can print.

Now pause here, because this is the disaster scenario, the exact shape of your worry. The company changed the model. The thing at the center of my system got worse, through no action of mine, with no vote and no warning. If my system had been built the way most people build on AI, wrapped around one model, tuned to its quirks, dependent on its goodwill, I'd have faced weeks of painful rework. At a company scale, the same event is somebody's very bad quarter.

Here's what actually happened. I opened OpenAI's Codex app, pointed it at my folder, the same folder, and went back to work. Same workflows. Same memory. Same logs, same instructions, same accumulated years of preferences. My video pages published that day, on schedule. I didn't change a single file in the system. Total migration effort: minutes, plus one small caveat I'll unpack below, honestly, because the caveat teaches the deepest lesson in the chapter.

The workflows never even knew.

That's not a lucky escape. That's a property. And by now you know exactly where the property comes from, because I've spent three chapters building it in front of you.

## Why the folder doesn't care

Run the 60-30-10 breakdown on the question "what actually depends on the AI company?"

The 60, the code: your fetch programs and parsers and publishers are ordinary programs. They don't know what AI is. A Python script runs identically whether it was cued by Claude, by Codex, or by an AI that won't be founded until next year. When my model turned over, every one of those programs kept working, and always will.

The 30, the memory: your instruction files, state files, and logs are text. Text is the most model-agnostic substance in computing. Any AI worth using reads and writes it fluently. My preferences didn't have to be re-taught, because they were never inside the model. They were in the notebooks, and the new conductor simply read them, the same way a human replacement would read their predecessor's files.

Which leaves the 10. The orchestrator's seat, cue, shape, judge, is the only chair in the whole system that an AI model actually occupies. And a seat is exactly what it is: a role, defined by your files, performable by anyone qualified. The model doesn't own the seat. It sits in it, at your pleasure.

You swap the driver, not the car.

That sentence is the whole chapter, so let me make sure it lands with its full weight. Everything durable about your system, the code, the accumulated memory, the structure, the conventions, the years of refinement, is the car, and you own it outright. The AI model is the driver: genuinely skilled, genuinely necessary, and genuinely replaceable on any given morning. When people fret about "betting on the wrong AI company," they're picturing systems where the company owns the car. In an AI-OS, no company owns anything of yours. The question "what if I pick wrong?" dissolves, because you never really pick. You hire, provisionally, forever.

And here's the inversion that should make you smile: the all-brain agent systems from Chapter 3 have this exactly backwards. When your system is 90 percent model, the model's quirks are your system. Its personality shifts become your process shifts; its price hikes are your rent. They built the car out of the driver. Meanwhile the design that uses the least AI turns out to be the design least at any AI company's mercy. Less dependence buys less dependence. It rhymes all the way down.

## The caveat that proves the rule

Now the honest caveat, because "I didn't change a single file" deserves a footnote, and the footnote is instructive.

Different AI apps look for their entry instructions in different places. Claude's tools read a file called `CLAUDE.md`. OpenAI's Codex reads one called `AGENTS.md`. Same idea, same contents, different filename on the front door. So the one real act of my migration was this: I had the new AI read the old instruction file and create its own version. A few minutes of conversation, one new file at the top of the folder, done. The substance, everything the file says, carried over untouched.

Picture the actual afternoon, because it was almost anticlimactic enough to be funny. I opened the new app, pointed it at the folder, and typed, more or less: "This folder is my AI Operating System. Read CLAUDE.md, that was written for a different AI, and create your own AGENTS.md equivalent. Then read the workflows directory and tell me what you find." A few minutes later it played the new hire on day one, impressively briefed: here are your workflows, here's what each does, here's what the logs show about recent activity. I had it run the research digest as a test. It read the instructions, ran the programs, checked the memory file, wrote the summaries, logged the run. First try. The entire "migration" that people build consulting practices around took less time than I've spent choosing a movie.

I include this partly for accuracy and partly because of what it reveals: the entire "lock-in" between my system and its former AI amounted to a filename. That's the moat. When people imagine switching AI providers, they imagine rebuilding; what I did was closer to reprinting the letterhead. And notice the detail that makes it so, because it's a design choice you'll inherit in Part IV: the instructions live in files, in the folder, in plain language, where any model can read them, rather than in some app's settings screen, where they'd have been marooned. Keep your system's soul in the folder and the front door can be renamed in an afternoon. (When you build yours, you'll simply create whichever file your chosen tool wants, and if you ever switch, you'll do what I did. Some people keep both files in the folder permanently, pointing to the same instructions. Perfectly legal.)

One more honest note while we're being honest: switching costs weren't zero because judgment work has flavor. Different models write a little differently, and for a workflow where prose style matters, my writing-heavy steps, I spent a short while nudging the new driver's tone, the way you'd calibrate any skilled new hire. The mechanical workflows didn't need even that. Days, not weeks, and shallow the whole way. Against "what if I built on the wrong company," it rounds to nothing.

## Freedom you'll actually use

It would be easy to file this chapter under insurance, valuable on the bad day, ignorable otherwise. But model-agnosticism turns out to be a tool you use constantly, not a policy you hope never to invoke. Three ways it shows up in ordinary weeks:

You stop dreading upgrades. Models now update the way phones do: often, and non-optionally. Each update lands differently on different work; the same release can improve your summaries and unsettle your formatting. Most people ride these waves with their fingers crossed. You have a lifeboat permanently tied to the dock, and knowing it's there changes the experience of every wave. Newest version acting up? Point the folder at the other app for a month. No drama, no forums, no petition. Your workflows won't notice.

And when you want to evaluate a change deliberately, whether a new version of your current model or a rival you're curious about, your system happens to be a ready-made test track. Here's the whole protocol, and notice it requires no expertise, just your files: pick two or three workflows whose output you know cold, ideally one mechanical and one writing-heavy. Run them with the new model. Put the results next to the last month of the old model's outputs, which your folder has been keeping without being asked. Read both like a customer: same instructions honored? Same formats? Prose you'd sign? Ten minutes, maybe twice a year, and you'll know more about how a model change actually affects your work than any benchmark chart or launch-day headline can tell you. Reviewers test models on puzzles. You get to test them on your Mondays.

You get to be a customer instead of a hostage. There are at least two world-class AI companies as I write this, plus a fast-improving field behind them, and your folder works with all of them. Not to mention open source local models that you can run on your own computer that are getting better every single day.

That's called a market, and you're on the good side of it. When prices or terms or quality shift, you can respond the way customers respond, by walking, which is precisely the customer the companies work hardest to keep. Hostages get pricing announcements. Customers get courted.

And, the fun one: your workflows themselves can hire specialists. Nothing says the AI orchestrating a workflow must be the AI that performs every judgment step. A workflow can reach out, mid-run, to a different AI that's better or cheaper for one particular job, the way a contractor brings in an electrician. I have a workflow that does exactly this, my orchestrator conducts, and at one step it hands a task to an entirely different company's AI, a specialist, for pennies. It's one of the five workflows I'm about to show you, so I'll leave the details for the next chapter, but see it for what it is: once no single model is load-bearing, models become a labor pool. You mix, match, and swap at will, in whole or in part.

The labor pool keeps deepening, too, in a direction worth a sentence even though I'll keep it hedged: beyond the big commercial models there's a fast-improving world of open models, ones you can run without paying anyone per token, sometimes on your own hardware. As I write this they're not what I'd put in my conductor's chair, and this book doesn't need you anywhere near them. But their existence is the floor under every price you'll ever be quoted: the companies renting you intelligence know a free-ish alternative keeps improving behind them. You benefit from that pressure without lifting a finger, and if the day comes when a workflow of yours could run its judgment slice on a free local model, your folder, need I say it, will not care.

There's a name for all this in business terms, and it's worth saying once without any metaphor: leverage. The structure of your system, one portable folder, mostly code and text, thin replaceable AI on top, is also your negotiating position with the most powerful technology companies on earth. You didn't have to be big to get it. You just had to keep the AI small.

## The tour is over

No assignment for this chapter. Instead, stand back and look at what you're now holding, because Part II is complete and you've earned the view.

Five chapters ago, "AI Operating System" was a phrase on a cover. Now: it's a way of working, not a product (Chapter 4). It lives in one folder you own, that moves anywhere and answers to no one (Chapter 5). Inside, a thin AI conducts, code labors, and the proportions have a slogan you could teach at a party (Chapter 6). The system remembers everything in files you can read, and compounds while chat windows reset (Chapter 7). And the intelligence at the center, the one piece you rent rather than own, is a driver you can replace any morning without the car noticing (this chapter). Every fear that brought you to this book has been met not with reassurance but with architecture: the job fear, the complexity fear, the cost fear, the lock-in fear, each one dissolved by a design choice you now understand.

What you haven't seen is what this machine can actually do when it's loose in a real business. You know my video-page workflow, but it's been carrying this book alone for seven chapters, and it has colleagues.

So next, the tour I'd give you if you visited: five real workflows from my real businesses, the money they save, the patterns they teach, and the one lesson that emerges when you see them side by side, which is that a workflow can be almost anything you can describe. Fair warning: somewhere in the middle of the next chapter, you're going to stop thinking about my workflows and start scribbling ideas for yours.

Bring the list.

# PART III: WHAT'S POSSIBLE

# CHAPTER NINE

## Workflows Worth Stealing

Eight chapters of concepts have earned you this: the tour.

This is the chapter where I open my own folder and show you what actually runs in there. Five real workflows, from my real businesses, with the real numbers attached. I picked these five out of the dozens I run because each one teaches a pattern you'll want for yourself, and because together they make the chapter's one big point, which I'll hold until the end, when you'll believe it.

The chapter title means what it says. Steal these. Not the specifics, you probably don't run YouTube channels, but the shapes. Every workflow here is a shape that fits a hundred businesses, and as you read, the exercise is to keep asking: what's my version of this?

First, a bit of context you need for the numbers to land.

That photo is my credibility slide, so let me say plainly what it shows: those plaques are YouTube's Silver Play Button awards, which the company sends when a channel crosses 100,000 subscribers. I have a stack of them because building and running YouTube channels, and the businesses attached to them, is what I do all day. I tell you this not to impress you but so you'll trust the arithmetic that follows. These aren't projections from a whiteboard. This is what my machine actually does, priced from my actual bills.

### Workflow one: the video page builder

You've known this workflow since Chapter 1. Time to see the whole thing.

Here's the strategy behind it. For every video I release on YouTube, my team builds a dedicated page for that video on the channel's own website. Not a lazy embed. A full page: a carefully written article that summarizes and complements the video, an extensive section of questions and answers drawn from it, and a detailed, formatted transcript where every section links both within the page and to the exact moment in the video itself. Click a line, jump to that second of footage.

Walk one of these pages top to bottom and there are six deliberate pieces: a heavily optimized title; the video itself, embedded; the article, written to stand on its own for a reader who never presses play; the Q&A section, which quietly matches the exact phrasings people type and speak into search; the transcript, formatted and linked so both humans and machines can navigate to any moment; and calls to action for the products connected to that channel. Six pieces, every page, every video, identically structured. Hold that "identically structured" in mind; it's about to matter when we talk about who builds it.

Why go to all that trouble? Because those pages get found. They're built for SEO, search engine optimization, the old craft of ranking in Google. And they're built just as deliberately for something newer: AEO, answer engine optimization. That term will be unfamiliar to most readers, so here's the plain version: more and more people don't search anymore, they ask, and an AI answers, in ChatGPT, in Perplexity, in Google's own AI summaries. AEO is the craft of being the source those AI answers draw on and cite. A clean article, a rich Q&A, a linkable transcript: that structure is exactly what answer engines love to pull from. As asking replaces searching, the pages keep working. The traffic those pages bring is a real, compounding asset for every channel I run.

Now the cost of that asset, the old way. Done by hand, one page takes about five hours of careful work after a video goes live. I release about one video per weekday on a given channel, so call it five pages a week. That's twenty-five hours of skilled human work, every week, forever. I used to pay a very good person eighteen dollars an hour to do this. Run the tape: about ninety dollars a page, about four hundred fifty dollars a week, about eighteen hundred dollars a month. Year after year.

Here's the workflow that replaced it. When a video goes live, I hand my AI the video's URL. That's my entire contribution: one link. A program fetches the video's details and complete transcript, the deterministic '60' doing the heavy lifting. The orchestrator, with everything on its desk, does the '10': writes the article, builds the Q&A, shapes the transcript with its section and timestamp links. Then programs again: assemble the page, upload it to the web server, publish. One instruction, about sixty seconds, and the token bill comes to three to five cents.

Ninety dollars and five hours, or a nickel and a minute. Same page. That's the entire thesis of this book in one before-and-after, and if you take nothing else from this chapter, take that.

But since this is the money chapter, let's follow the money one step further, because "eighteen hundred dollars a month" is one of those figures that slides past too easily. Suppose you just kept paying it: over ten years, that's $216,000 out the door. Now suppose instead the task is automated and you invest the same eighteen hundred every month as it frees up. At a 15 percent annual return, ten years of those monthly contributions grows to roughly $495,000, call it half a million: your $216,000 in contributions plus about $279,000 in growth. I use 15 percent because that's what my own index-fund option strategy has returned over the years, and I want to be crystal clear: that is a description of my portfolio, not advice for yours, I'm not a financial advisor, and returns like that are far from guaranteed. So run it conservatively instead: at 8 percent, the same monthly investing lands around $330,000. The exact endpoint isn't the point. The point is the fork in the road: one path, the money is simply gone; the other, one automated task quietly becomes a few hundred thousand dollars. And that's one workflow. (In reality wages rise, so the do-nothing path costs even more than the flat math suggests.)

The shape to steal: find the recurring task you're paying real money for, where most of the work is mechanical and the judgment slice is thin. That's the page builder's shape, and versions of it hide in product listings, client reports, property descriptions, podcast show notes, newsletter issues. Yours is in there somewhere.

## Workflow two: the writing workflow

Now forget everything impressive about workflow one, because the second workflow exists to correct the impression the first one leaves.

Years ago a book changed how I write: *On Writing Well* by William Zinsser. I've written many books over the years, and that one shaped them more than any other. When I started building my AI-OS, I did something with it that still feels a little like a cheat code: I gave the book to my AI and had it study how Zinsser writes, then distill what it learned into a workflow called, simply, "writing."

Now, any time my AI writes anything for me, a blog post, a video script, an email that matters, I run the draft through the writing workflow, and out comes prose with that discipline applied. Every time, on command, forever.

Look inside this workflow's folder and here's what you'll find: almost nothing. No programs to speak of. No database, no API connections, no publishing machinery. It's essentially one instructions file holding a distillation of writing principles, plus the standard log. It is the smallest workflow I own, and I use it more than almost any other. (I warned you in Chapter 6 that you'd meet a workflow proportioned nothing like 60-30-10. This is it, nearly all judgment, and none the worse for it.)

That's the lesson, and it's why this workflow follows the flashy one: a workflow doesn't have to be a machine. Workflow one is a complex contraption with a dozen moving parts. This is a single page of distilled taste. Both are workflows. Both earn their keep weekly.

And notice the deeper move, because it's the part worth stealing: feed the AI a body of expertise once, distill it into a workflow, and you've turned something you love into a tool you can invoke forever. Extract once, use forever. Your version might not be a writing book. It might be your own best proposals, distilled into a proposal workflow. Your most successful listings. The way your best-ever employee handled complaints. Expertise you already own, sitting in documents, waiting to become a button.

Think about what you'd distill, and don't limit yourself to documents you already have. The extraction conversation itself can create the document: sit with your AI for an hour and talk through how you decide what you decide, how you price a job, what makes you turn down a client, what your eye checks before anything ships, and have it write the distillation as you talk. People pay for this under grander names, "capturing institutional knowledge," "founder's playbook", and it's sitting in your chat window for the cost of an honest hour. The workflow at the end is almost a bonus. The document is your judgment, finally outside your head, which is worth having even on the days no AI reads it.

I'll add the honest disclosure, since this book preaches receipts: the words you're reading went through that workflow. I drafted, and the drafts ran through the same writing pass I just described. You're holding its output. I can't offer a cleaner proof that these things work than that.

### Workflow three: the thumbnail maker

This one settles a promise from Chapter 8, where I told you a workflow could hire a specialist from a different AI company mid-run.

Every YouTube video needs a thumbnail, the little billboard image that makes someone click, and thumbnails are serious business. Channels live and die by them. A good freelance thumbnail designer charges twenty-five to seventy-five dollars per image, and at the volume I publish, that adds up fast.

Here's my version now. I give the workflow one thing: the video's title. The workflow takes it from there: it connects to Google's Gemini image AI (a model that, as I write this, goes by the unlikely name Nano Banana), my current favorite for this job, though as you now well know, favorites rotate, and generates a professional thumbnail featuring my actual face, in the exact visual style my channels use. The style rules and reference images live in the workflow's instruction files; the face is mine; the output is consistent enough that my thumbnails look like a family, which is exactly what a channel wants.

Cost per thumbnail: three to six cents. Against twenty-five to seventy-five dollars for the human version. The ratio is so lopsided it reads like a typo, but there it is again, the same thousand-to-one gap as workflow one, in a completely different craft.

Notice the architecture, because this is the pattern to steal: my orchestrator, the AI conducting the workflow, is one company's model. The image specialist is another company's entirely. The workflow reaches out, mid-run, hands the specialist a job, takes the result, and moves on. This is what I meant when I said that once no single AI is load-bearing, the models become a labor pool. Your AI-OS isn't a one-AI system; it's a hub that can call whatever specialist the task deserves: image AIs, transcription AIs, translation AIs, whoever's best and cheapest this year.

One practical note, kept short because Part IV handles it properly: connecting to an outside service like this involves an API key, the password-like credential from Chapter 5, and this workflow is exactly why the rule there matters. The key lives on my machine, not in the folder. When we build yours, the AI will set this up with you in a few minutes.

### Workflow four: the competitive research summarizer

You've been hearing about this one all book, it's the "research digest" from my Chapter 4 morning, and I saved the full reveal for here because it's the payoff of the memory chapter.

The problem it solves: in my line of work I need to know what other channels in my niches are doing, what's landing, what topics are surfacing, without spending my week watching YouTube. Multiply a list of channels by daily uploads by ten-to-twenty minutes a video and monitoring the competition is a part-time job.

The workflow: I keep a list of channels worth watching. A program checks every channel on the list and compares what it finds against the workflow's memory file, the list of every video it has already processed. Chapter 7's trick, exactly: new equals world minus notebook. For each genuinely new video, a program pulls the transcript, and the orchestrator writes two things: a two-or-three-sentence summary, the quick glance, and a fuller one-paragraph summary, the deeper read. Title, both summaries, and a link land in my morning digest.

A single entry in the digest looks something like this: a channel name and video title, then the glance, "Tests five budget microphones against a $400 standard; the $89 one wins for spoken voice. Strong comment reaction to the blind-test format.", then, below it, a paragraph with the substance, and finally the link. Multiply by however many new videos appeared across the watch list, and that's the whole document. Nothing fancy. The value isn't in any entry; it's in what the document quietly proves: every channel got checked, every new video got watched, by something with infinite patience, and I read only the residue.

The design detail I'm proudest of is the tiering, and it's worth stealing on its own. The digest respects my attention: I skim titles and short summaries in a minute or two; where something catches, I read the paragraph; only if I'm still curious do I click through and spend real minutes on the video itself. Most days I never click at all. The workflow's job isn't to feed me information; it's to let me decide, three times over, how much information to want.

And here's memory earning its keep, live: the reason the digest contains only what's new to me, no repeats, no noise, is one text file listing what's been seen. Delete that file and the workflow becomes a firehose. That single file is the difference between an assistant and an annoyance, and it's the concrete answer to anyone who wonders whether the memory chapter was really necessary.

The shape to steal is monitoring anything: your competitors' prices, your industry's news, your supplier's catalog, your own reviews, mentions of your business. If you've ever wished someone would just check and tell you what changed, this is that someone, at three to five pennies a morning. And workflows like this run on a schedule beautifully; mine fires with my coffee, and in Part IV I'll show you how to make one fire without you.

### Workflow five: US Metrics

The last workflow exists partly for what it does and partly to break a spell, because the first four all came from my media business, and you might be quietly concluding this is all YouTuber stuff.

So: this one has nothing to do with content, channels, or anything I sell. I'm simply a person who likes to know how the economy is doing, without wading through data tables or waiting for someone's spin on cable news.

The Federal Reserve Bank of St. Louis runs a free public service called FRED, an enormous library of US economic data, unemployment, inflation, interest rates, hundreds of thousands of series, available to anyone through a straightforward connection. Every week, my "US

Metrics" workflow queries FRED for the handful of indicators I care about, a program does the fetching and assembling, and the orchestrator turns the numbers into a short, readable report: what moved, what didn't, what's drifting somewhere interesting. It shows up weekly. I read it in two minutes, and I walk around with a current, first-hand sense of the economy that most people I talk to don't have.

The shape to steal: there is public data about whatever you care about, your industry, your region, your hobby, your kids' school district, and most of it sits in sources nobody has the patience to check. A workflow with patience is cheap. Yours could watch housing permits in your county, fuel prices on your trucking routes, enrollment trends, weather patterns for your farm, anything with a source. The judgment slice, the part worth an AI, is the translation: numbers into "here's what this means this week, for you."

## A workflow can be anything

Now stack the five side by side and let the range register. A publishing machine that replaced a payroll line. A single page of distilled taste. An art department for six cents. A tireless scout with a perfect memory. An economist on retainer for pennies. One of them is all code and a sliver of AI; one is nearly all AI and no code; the others sit everywhere in between. Some publish, some watch, some create, some translate.

Here's the chapter's one big point, the one I've been holding: there is no category called "things that can be workflows." A workflow is any task you can break into describable pieces. That's the entire membership requirement. If you can say what comes in, what should go out, and roughly what happens in the middle, it can probably be a workflow, whatever your business, whatever the task, as long as it's something done on a computer or online. (That's the honest boundary: my AI-OS cannot stock a shelf or drive a delivery route. Yet.)

To stretch your sense of the range, here's a lightning round of shapes I've either built, seen built, or sketched with people from other industries, one line each, offered as sparks rather than instructions. A review responder that drafts replies to new customer reviews in your voice, kind ones and rough ones, for your approving click. A no-show reducer that drafts tomorrow's appointment reminders with each client's history in mind. An inbox triager that reads the morning's email and hands you a ranked list of what actually needs you. A listing builder that turns photos and a voice memo into a formatted product or property listing. A grant watcher that checks funding databases weekly for anything matching your nonprofit. An invoice chaser that knows who's late, how late, and drafts the escalating nudges. A meeting distiller that turns a recording's transcript into decisions, owners, and deadlines. A social repurposer that turns each blog post into platform-shaped fragments. A price-list differ that reads supplier PDFs and tells you what quietly went up. None of these took programming to describe; each is an afternoon with the method you now have; and the list's real point is what it does to your own: somewhere in those nine, you probably heard your business clear its throat.

And when you're not sure whether some task qualifies, or how to break it into pieces? Ask your AI. I mean that literally; it's maybe the most practical sentence in this chapter. Describe the task to your AI in plain words and ask how it would structure the thing as a workflow. You'll be surprised how often it finds an angle you didn't see. The skill, as ever, isn't knowing how to build. It's describing what you want, and by now you've had eight chapters of practice.

So, the assignment, and this week it's the real one. Take out your list, the sentences you've been collecting since Chapter 1, the one you sorted into columns in Chapter 6. Read it once more, next to this chapter's five shapes. Circle the one that's most like workflow one: recurring, mostly mechanical, costing you real time or money. That circle is your first build.

Because the tour is over, and the concepts are over, and in the next chapter, you and I are going to sit down with your AI and build the thing.

* * *

*The five workflows in this chapter are the short versions. I teach the full builds, over the shoulder, screen by screen, in the AI Operating System courses at SectionZero.ai, along with a growing library of workflows that didn't fit in a book. If watching one built end to end sounds useful, that's where to look.*

# PART IV: BUILDING IT

# CHAPTER TEN

## Your First Workflow, From Scratch

Every skill has a moment where reading about it ends and doing it begins. Golf has the first tee. Cooking has the first pan that's actually hot. This chapter is that moment for you, and I want to tell you up front what it's going to feel like, because managing your expectations is half of getting this right.

Your first build will take about an afternoon, and most of that time won't be work; it'll be the pleasant, slightly surreal experience of describing a chore to a machine and watching the machine take it seriously. There will be a moment, probably around the first successful run, where you laugh out loud. There will also probably be a moment where something comes out wrong and you feel a flicker of "I knew this was too good to be true." Both moments are part of the recipe. By the end you'll have a real workflow, doing a real task from your real life, and, more valuable than any single workflow, you'll have the rhythm of building, which is the thing this chapter actually teaches.

I'm going to walk the whole road with you: setting up, the first conversation, the build, the first run, the fix, and the run-it-again that makes it permanent. So this chapter follows one continuous example, and I've chosen the task you've already done the thinking for: the Monday summary you sorted into columns in Chapter 6 and checked against notebooks in Chapter 7. Where your circled task differs, and it will, the shape of everything stays the same.

## The ten-minute setup

First, the tools, and the two honest costs.

You need an AI app that can do three things: hold a conversation, read and write files in a folder you point it at, and write and run small programs. As I write this, the two I use are Anthropic's Claude desktop app (Cowork or ClaudeCode) and OpenAI's Codex app. Download either from its official site the way you'd download anything. If you're reading this a few years on and different names dominate, you know by now that it doesn't matter: any tool of this caliber works, and Chapter 8 taught you exactly how little you're committing to by picking one.

Honest cost number one: the good tiers of these apps are paid. Plans change too fast to print, but as I write this you should expect something like the price of a couple of streaming subscriptions per month ($20 - $100 based on usage). Weigh that against Chapter 9's arithmetic and it's the best subscription your business will ever carry, but I want you walking in knowing it exists. There is no other bill. Remember Chapter 3: the whole design exists to keep the metered part small, and your workflows will cost pennies to run and likely be included in the cost of the subscription that powers your conversations.

Honest cost number two: an afternoon of your attention, today, and less each time after.

Now the setup itself, which barely deserves the word. Open the app. Sign in. Point it at the folder you created at the end of Chapter 5, there's a setting or a button in every such app for choosing the folder it works in; if you can't find it, ask the app itself, it knows. That's the entire installation. No terminal, no configuration files, nothing to compile. If your folder lives in Dropbox like we discussed, you're building portable from minute one.

You're looking at a chat window pointed at an empty folder. Here we go.

## The first conversation

You start by telling the AI what this folder is going to be. Not a magic incantation, remember the rule from my system that there are no exact prompts, just a plain statement of intent. Yours can be a cousin of this:

"This folder is going to be my AI Operating System, a place where you and I build automated workflows for my recurring tasks. Set up the basic structure: an instructions file for yourself at the top level, and a workflows folder where each workflow will live in its own subfolder. Keep the top-level instructions lean. Each workflow will have its own instructions file (context.md), any code it needs, its own memory files, and a log. Three rules for everything we build: follow workflow instructions exactly as written, don't improvise; if anything goes wrong mid-run, stop, log the error, and ask me; and when creating workflows, have them use as much deterministic code to do the thing as possible (Python, Javascript, whatever seems best for the task), you as the AI should be the 10% orchestrator not the heavy lifter in the workflow. Sound good?"

Send it, and watch your empty folder sprout. The AI will create its own entry instructions file at the top, `CLAUDE.md` or `AGENTS.md` depending on your app, plus the `workflows` folder, and it will tell you what it did. It may suggest additions. Fine; this is a conversation, and its suggestions are usually good. What you've just done, in one message, is turn a folder into an AI-OS: you've given the conductor its standing orders.

Open the instructions file it wrote, remember, it's plain text, and read it. This is a habit worth starting on day one, not because you need to check its work, but because reading that file is how the system stays yours. It should say, in the AI's own wording, roughly what you told it. Nothing mystical. A memo, from your system, to itself.

Now the workflow. Here's where your Chapter 6 worksheet pays off, because a good build request is just your three columns read aloud. Something like:

"Let's build my first workflow. Call it weekly-summary. Every Monday I collect last week's sales numbers and email a summary to my business partner. Here's how it works today: the numbers live in a spreadsheet my point-of-sale system exports; I'll drop that file into the workflow folder each Monday. The summary compares this week to last week, and calls out anything unusual, a big jump, a big drop, anything weird. My partner likes it short: a few lines, numbers up top, plain talk. Set this up as a workflow I can run every Monday, and keep a memory of past weeks so the comparisons get smarter over time."

Notice what you did and didn't specify. You said what comes in, what goes out, and what judgment looks like ("anything weird," "plain talk"). You didn't say a word about how, no file formats, no programming languages, no mention of code at all. The how is the conductor's job, and it will now show you something that never stops being satisfying: the AI thinks out loud, in plain English, about how to structure your chore.

Expect questions back. Good ones. Mine asked, when I built my version of this years ago, what "unusual" meant to me, did I want a fixed threshold or its judgment? (Its judgment, I said, but tell me the number that triggered the comment.) Yours might ask what to do if the spreadsheet is missing a day, or whether "email a summary" means it should draft the email for you to send. Answer honestly, and notice the questions are the AI finding the edges of your task, the exact places where un-articulated assumptions live. This interview is the AI doing you a professional courtesy that human contractors charge for.

Then it will likely propose a plan, and the plan will be Chapter 6 wearing work clothes. Here's roughly what mine sound like, so you recognize the species: "Here's how I'll structure this. I'll write a small program that reads the export and computes totals, comparisons to previous weeks, and anything crossing the thresholds we discussed; that part should be code, not AI, so it's identical every week. I'll keep your formatting preferences and the definition of 'unusual' in the workflow's context file, and a running memory of each week's numbers so comparisons deepen over time. When the program finishes, I'll write the summary in your partner's preferred style, flag anything the numbers can't explain, and log the run. To build this I need one sample export. Want me to proceed?" Read that twice, because it's the book so far, recited back to you by the thing that's about to do the work: doing, knowing, deciding, receipts. You say yes. It asks for the sample spreadsheet. Drag one in. It builds.

While it builds, the app narrates, and the narration is worth watching your first time, not because you need to supervise, but because it dissolves the last of the mystery. You'll see it announce each move in plain English: creating the folder, writing the context file, writing the program, testing the program against your sample, adjusting, done. What you're watching is the answer to "but how does the code get there?", the question that quietly stops most people from ever starting. It gets there the way the memo gets there: the AI types it. You watched.

Two or three minutes later, your folder contains a new subfolder called `weekly-summary`, and inside it: an instructions file, a small program you'll never read, a memory file, and a fresh log. Total elapsed time since you opened the app, maybe twenty minutes.

You've built software. Sorry, there's no other honest verb for it.

### The first run, and the flicker

"Run the weekly-summary workflow."

Here's what a first run feels like from your chair: a few seconds of the AI narrating its own progress, reading its instructions, running the program, examining what came out, then the summary appears in the window. Numbers up top, a few lines of plain talk, a note that Thursday looked odd.

Read it the way your partner would. And now, the moment I promised: something will be not-quite-right. Maybe it counted a refund as a sale. Maybe "short" came out as three paragraphs. Maybe Thursday wasn't actually odd, your point-of-sale exports Thursdays weird, and you've known that for years, but nobody told the workflow, because the person who knew it, you, forgot it was knowledge.

This is the flicker, the "too good to be true" moment, and I need you to reframe it right now, permanently: nothing failed. What happened is that your description of the task met reality, and reality pointed out what your description left out. This happens with every new employee you've ever trained, and you never once concluded that employees don't work. You said: oh, right, and told them about Thursdays.

So tell it about Thursdays: "Good first run. Two fixes: refunds shouldn't count as sales, they're in the numbers as negatives. And Thursday's pattern is normal, the export does that every week; only flag Thursday if it's unusual even for a Thursday. Add those changes to the workflow instructions."

And here's the difference between this correction and every correction you've ever typed into a chat window, the difference this whole book has been building toward: watch what the AI does with it. It doesn't just apologize and adjust. It edits the workflow, the refund rule goes into the program, where it's now deterministic law; the Thursday wisdom goes into the instructions or memory, where every future Monday will read it. Say it once, own it forever. Dana's rentals, your Thursdays.

Run it again. Better. Maybe perfect; maybe one more round, and the second round is usually finer-grained than the first: less "that's wrong" and more "closer, but my partner would want the comparison as a percentage, not a raw number." Notice your own corrections getting more precise as the output gets more right; that's not fussiness, that's the process working, because vague dissatisfaction can't be written into a file and specific preferences can. Two or three of these loops is completely normal for a first build, each one takes about a minute, and each one is doing something subtle: it's transferring your taste from your head into the folder. That taste transfer is the actual work of the afternoon. The code was never the hard part.

A note on the feeling of those loops, because it trips people who are used to software: correcting a workflow doesn't punish you the way changing your mind usually does. There's no settings maze to re-navigate, no fear of breaking what already works, no starting over. You say the sentence; the right file absorbs it; everything else stands. This does something to your willingness to have standards. People tolerate mediocre output from tools because demanding better costs too much fuss. Here, demanding better costs one sentence, so demand better. The workflow you'll be running in a year is made of every sentence you bothered to say in week one.

## The Monday after

Now comes the part that justifies everything, and it's the least dramatic moment in the chapter. Next Monday, you open the app, drop in the new export, and type: "Run my weekly summary."

That's it. That's the whole interaction, this Monday and every Monday, forever. No re-explaining, no context, no reminding it about refunds or Thursdays. The folder remembers everything, the program crunches, the memory deepens by one more week, which means next week's "compared to usual" is one week smarter, the log quietly notes another successful run, and the summary lands, in your partner's preferred flavor of short.

Thirty minutes of Monday, converted into about ninety seconds, for the price of one afternoon and a couple of corrections. And I want to point at what you actually own now, because it's more than a time-saver. Open the folder and look: instructions in readable English, a memory accumulating your business's weekly rhythm, a log of everything that's ever happened, and a program that will do the arithmetic identically until the sun burns out. Portable, inspectable, yours. If the AI model you use vanishes tonight, you point a different conductor at the folder tomorrow, Chapter 8 lives in this folder too.

One more thing before the assignment, because first builds occasionally hit a bump bigger than a correction, and I want you equipped, not surprised. Sometimes a run just breaks: the program errors, the file won't read, something. Your workflow already knows what to do, you gave it the rule in your very first message: stop, log it, ask. So it will halt and ask you, and your entire job is to have the conversation: "It failed, here's the log, figure out what went wrong and propose a fix." You are never the mechanic. You're the person who says "it's making that noise again" while the mechanic listens, and in this shop, the mechanic wrote the engine and reads its own logs. In my time running dozens of workflows, I have never once needed to read the code to get something fixed. The conversation is always enough.

Here's what such a moment actually looks like, so your first one lands as recognition instead of alarm. Third Monday, say, you drop in the export and run it, and instead of a summary you get: "The run stopped at the reading step. The export has a different column layout than previous weeks, there's a new column between order totals and dates. I've logged the details. Should I update the program to handle both layouts, or is the new one permanent?" Your point-of-sale updated itself over the weekend; the world shifted, the way worlds do. Read that message again, though, and inventory what didn't happen: nothing plowed ahead, nothing guessed, no hollow summary went to your partner, and the machine arrived not just with the problem but with a diagnosis and two sensible options. You type "new one's permanent, update it." Ninety seconds later you have your summary. That's a breakdown, in this system. If you've ever spent an afternoon on hold with a software vendor over less, you understand what you're being handed.

And if a first build ever goes truly sideways, confusing on top of broken, here's the escape hatch that costs almost nothing: delete the workflow folder and build it again. That sounds drastic only if you're used to software where setup was expensive. Your investment here was a twenty-minute conversation, and the second conversation is always better, because you've seen the failure and can describe the task around it. I've rebuilt workflows rather than untangle them, deliberately, the way you'd rewrite a confusing paragraph instead of patching it. The folder is sand, not marble. Sculpt accordingly.

## Now yours

The assignment writes itself: build the circled one. The task you ringed at the end of Chapter 9, this afternoon if you can, this week at the latest, while the rhythm of this chapter is still in your ears.

Here's the whole recipe on one card, portable to any task:

Tell the AI what the folder is, once, with the three rules: follow instructions exactly; stop and ask on errors, use deterministic code when possible in workflows. Then describe your task from the worksheet: what comes in, what goes out, what judgment looks like. Answer its questions honestly; drag in a sample of whatever it asks for. Let it build. Run it. Correct it like a promising new hire, once, twice, three times, knowing every correction is permanent. Then run it for real, and notice, sometime next week, the moment you type "run my..." and realize you're not doing that chore anymore.

A word about ambition, from someone who's watched people start: resist the urge to make your first build the big one. Not the client-facing thing, not the one with four data sources and a deadline. Pick the boring, honest, weekly chore. First workflows are for learning the rhythm, and the rhythm transfers; the stakes don't need to. Your fifth workflow can be ambitious. Your first should be almost embarrassingly small. Mine was.

And when it works, when you run it the second Monday and it just... works, do one thing for me: open the folder and look around. A week ago this was empty, and everything now in it came out of sentences you typed. The instructions file, the memory, the log, even the program: all of it precipitated out of a conversation, like the whole thing had been dissolved in plain English all along.

That folder full of files raises a reasonable question, though. What exactly is all that stuff? You've been trusting the tour guide's summary, an instructions file, a memory, a log, but you're an owner now, and owners get to see the deed. In the next chapter we open every file in a working AI-OS, including a few I haven't told you about yet, one of which is where your AI keeps its personality, and read what's actually written inside.

# CHAPTER ELEVEN

## Inside the Folder

Some people buy a house and never open the electrical panel. Nothing wrong with them; the lights work anyway. But there's a particular kind of owner who, on the first weekend, walks the whole property, opens every panel, finds the water shutoff, and learns which switch does what. Not because they plan to rewire anything. Because it's theirs, and owning a thing you've never looked at feels like renting with extra steps.

You built a workflow last chapter. The lights work. This chapter is the walk-through for that second kind of owner, which, ten chapters in, is what you've become. We're going to open a real AI Operating System, mine, in its structure, and look at every kind of file in it: what it's called, what's written inside, and why it's there. None of this is required reading for your workflows to run. All of it is required for the feeling I want you to leave this book with, which is not "I have a clever machine" but "I know my machine."

Here's the whole estate at a glance, and then we'll walk it:

```
my-ai-os/
├── bootstrap.md          <- startup instructions: "read this first,
every time"
├── CLAUDE.md (or AGENTS.md)  <- the entry point: what this place is,
how to behave
├── soul.md               <- optional: your AI's personality
├── principal.md          <- optional: who you are, what you care about
└── workflows/
    ├── _template/        <- the scaffold every new workflow is built
from
    ├── weekly-summary/
    │   ├── context.md    <- this workflow's instructions
```

```
│   ├── (a small program or two)
│   ├── (memory files, format fits the job)
│   └── logs.md       <- every run, recorded
└── ...one folder per workflow...
```

Top floor first.

## The front door: your entry file

You met `CLAUDE.md` (or `AGENTS.md`, depending on your app) in the last chapter; the AI wrote yours in the first five minutes. It's the file every AI tool of this kind reads on arrival, the front-door memo that answers: what is this place, and how do I behave here?

Open a good one and it's almost disappointingly short. Mine says, in compressed form: this folder is an AI Operating System; workflows live in `workflows/`, one folder per task; each workflow's own `context.md` is the authority on how it runs; follow workflow instructions exactly, without improvising; on any error, stop, log it, and ask; log the start and end of every run; when building a new workflow, copy the template, use as much deterministic code as possible to achieve the workflow's purpose. Plus a line or two of housekeeping. That's the whole constitution.

Since prose descriptions of files only get you so far, here's what that actually looks like on the page, condensed but true to form:

> This folder is an AI Operating System. Workflows live in `workflows/`, one folder per workflow, one task per workflow.
>
> Each workflow's `context.md` is the authority on how that workflow runs. Follow workflow instructions exactly as written. Do not improvise or optimize. If you see a better way, propose it and wait.

> On any error or unexpected condition: stop, record it in that workflow's `logs.md`, and ask. Never work around a problem silently.
>
> Push as much of the workflow as possible into deterministic code, leaving AI to handle only a thin (roughly 10%) orchestration role.
>
> Log the start and end of every run in the workflow's `logs.md`.
>
> To create a new workflow: copy `workflows/_template/`, name the folder for its single task, and fill it in.

Remember, before you're tempted to copy that verbatim: yours will be worded differently, because your AI will write it in its own phrasing during your first conversation, and Chapter 12 will explain why that's not just acceptable but correct. I show it to you so the words "entry file" stop being abstract, not to hand you scripture.

Why so short? This is the caveat I want to hand you early, because it's the one thing in this chapter that will genuinely protect you over the years: the entry file wants to grow, and you must not let it. Every month you'll be tempted to add one more instruction, one more preference, one more special case. Resist. This file gets loaded into the AI's working attention every single session, and the bigger it gets, the more the AI is juggling, and the likelier it is to fumble, to miss instructions or, worse, to start improvising around them. The design defends itself the same way this whole book does: push detail down. Global rules live here, briefly. Task detail lives in each workflow's `context.md`, which the AI reads only when running that workflow. Keep the front door a doorway, not a warehouse.

## The odd little file that makes it all reliable

Next to the entry file sits the smallest and, pound for pound, most important file in my system: `bootstrap.md`.

Here's the awkward truth it exists to fix. The AI apps are supposed to read their entry file automatically when they start (either CLAUDE.md or AGGENTS.md). In practice, they usually do, and sometimes don't. Sessions start in odd states; apps update; attention wanders (theirs, not yours). And an orchestrator that skipped the constitution is exactly the improvising, non-logging, plow-ahead-on-errors AI that this whole design exists to prevent.

The fix is almost comically low-tech: a tiny file of startup instructions, and a habit. The file says, in essence: read the entry file and follow it; then do any other startup ritual we've agreed on. The habit is that every time I sit down to work, my first message is the same five words: "Read bootstrap.md and follow it."

That's it. That's the whole trick. It costs three seconds, and it guarantees the session starts with the standing orders actually standing. Belt, meet suspenders: the app will probably read the entry file on its own, and if it does, no harm done, the bootstrap just confirms it. If it doesn't, the bootstrap catches it. You'll develop the habit within a week, it becomes as automatic as a turn signal, and it's a nice bit of symbolism besides: every working session begins with you, deliberately, handing the conductor its baton.

## The two fun ones

Two more files can sit at the top level, and neither is necessary, which is exactly why I like telling you about them. A system with only load-bearing parts is a machine; the optional parts are where it becomes yours.

`soul.md` is your AI's personality file. Whatever identity, voice, or character you want your orchestrator to have, you write it down here, and the entry file points to it. Some people keep it strictly professional: brisk, no flattery, no exclamation points. Some name their AI and give it a whole disposition. Mine has one, and I'll admit the file is longer than it strictly needs to be. It changes nothing about how the code runs. It changes a surprising amount about how the working day feels, and you spend your working day with this thing.

`principal.md` is the mirror image: a file about you. The principal, the person this system serves. What you do, what you're building, how you like to communicate, what you chronically forget, what "good" looks like to you. The orchestrator reads it and stops being a brilliant stranger; it's context that makes a hundred small judgment calls come out closer to how you'd have made them, without you specifying anything workflow by workflow.

Both files are text, both are read-on-arrival, and both grow slowly over time. Skip them on day one if you like. You won't skip them forever; nobody does.

If you want a feel for what goes in them, a few lines from the genre. From a `soul.md`: "Be direct. Skip the flattery; never open with 'Great question.' When I'm about to do something expensive or irreversible, get skeptical on my behalf. Dry humor welcome." From a `principal.md`: "I run a twelve-table restaurant and I'm not technical; explain things like I'm smart but busy. Mornings are prep time, so keep anything before noon short. I will always choose simple-and-reliable over clever. Pet peeve: being shown five options when one is clearly right." None of this is configuration in any technical sense; no setting anywhere changes. But every judgment call in every workflow gets made by something that has read those lines, and the cumulative effect is the difference between staff who know you and staff who are new every day, which, after the memory chapter, you'll recognize as the entire theme of this book applied to personality.

If you create these files, tell your AI to read them in the bootstrap.md file. That way, every time you start the app and tell your AI to read your bootstrap.md file, it will be instructed to read the soul.md and principal.md file (as well as the CLAUDE.md or AGENTS.md file).

## The workflows floor

Down one level, the `workflows/` directory: one subfolder per workflow, and the rule you've absorbed by osmosis is worth stating as law once. One workflow does one task. Never more.

You'll be tempted, someday soon, to build the mega-workflow, the one that does the Monday summary and files the invoices and checks the competitors, because they all happen on Monday. Don't. Six small workflows beat one large one, every time. Small workflows are easier to describe (one sentence each), easier to fix (when something breaks, you know which room it's in), easier to trust (each has its own log and memory), and easier to reuse and share. This is the folder-level echo of the same principle that runs the whole book: many small deterministic parts, thin coordination on top. Your AI-OS is itself a 60-30-10.

Inside each workflow folder, the same four-part household you built in Chapter 10, and here it's worth showing a `context.md`'s actual bones, because this file is the one you'll read most often. A typical one runs in sections, something like: **Purpose** ("Produces the Monday sales summary for my partner from the weekly point-of-sale export"). **How it's invoked** ("John drops the export into this folder and says 'run the weekly summary'"). **Steps** (numbered: log the start; run the program on the export; review its output against the rules below; write the summary; log the end). **Rules** (the summary format and length; refunds aren't sales; the Thursday clause; and the two commandments again, follow exactly, stop-and-ask on errors). **Memory** (what's stored and where). **Files** (what each file in this folder is). One page, maybe a page and a half, all sentences. A stranger, or a brand-new AI model, could pick up this workflow cold from that one file, which is precisely the test of a good one.

So, formally: `context.md`: the workflow's own instructions, what it's for, how it's invoked, the steps, the rules, and, in its own file-level copy, the two commandments (follow exactly; stop and ask). Yes, they're repeated here even though the entry file says them globally; instructions get followed better when they're written where the work happens, true of AIs and, you may have noticed, of humans. The supporting programs: whatever small code the AI wrote for the mechanical steps. The memory: whatever the job needs, a Markdown list, a data table, sometimes a real database file when a workflow tracks thousands of things; the AI picks the container to fit the contents, and it all lives right here, moving when the folder moves. And `logs.md`: start and end of every run, plus anything that went wrong. Memory is what the workflow needs to know; the log is what it did. You knew that. It's on the wall of Chapter 7.

## The template: your system's DNA

One resident of `workflows/` isn't a workflow at all, and it might be my favorite structural trick in the whole system: a folder called `_template/`.

Inside is a blank workflow. A fill-in-the-blanks `context.md`, purpose here, invocation here, steps here, with the logging steps pre-written and the rules block already in place, and a starter `logs.md` with the entry format spelled out. And in the entry file upstairs, one standing instruction: when we build a new workflow, copy the template and fill it in.

Think about what this does. Every future workflow is born already shaped: logging built in, error rule in place, structure consistent with every sibling. The AI never reinvents the pattern, never forgets the log file, never freelances the format, because it doesn't start from a blank page; it starts from your known-good bones. When people ask me why my dozens of workflows all behave like a family, this is the answer: they're all the same species. The template is the DNA.

You don't have to design one. Tell your AI, "based on the workflow we just built, create a `_template` folder so future workflows follow the same structure," and it will, in about a minute. Do it right after your first build, while the shape is fresh. Your second workflow, and your fiftieth, will be built from it.

### The one thing that must never live here

Now the safety briefing. Chapter 5 gave you the principle; the thumbnail workflow in Chapter 9 made it real; here's the mechanics, and this is the single most important operational rule in the book, so it gets its own section.

Some workflows connect to outside services, an image AI, a data source, your email provider, and those connections are unlocked by an API key: a long string of characters that works like a password. Here's what makes API keys dangerous in a way regular passwords aren't: they don't ask twice, and they spend money silently. A person who has your key is you, as far as that service is concerned, no login screen, no verification code, and if the service bills by use, they're you on your credit card.

Your folder, meanwhile, is designed to travel: it syncs to Dropbox, it snapshots to GitHub, someday you might share a workflow with a colleague. Folder that travels plus key that spends equals the one genuinely bad day this system can produce. So the rule: keys live on the machine, never in the folder.

The standard mechanism is called an environment variable, which sounds technical and is just a named note that lives with your computer's account settings rather than in any folder: the key sits there, the workflow says "read the key from the note named such-and-such," and the folder itself never contains the secret. It syncs, snapshots, and shares clean. You will not set this up by hand; when a workflow first needs a key, tell the AI to walk you through storing it as an environment variable, and it will, step by step, in a few minutes, for your exact computer.

(You may see an alternative, a `.env` file inside the folder, mentioned by your AI or in things you read; it works, and it's the thing most likely to leak the day the folder syncs somewhere public. If a key has to be in a file for some reason, that file must never travel. Given the choice, and you're always given the choice, take the environment variable.)

One rule. It's the whole briefing. Everything else in this folder can be shared with the world.

## There is no correct folder

The tour's done, and before we climb back out, the point that makes it all land, the one I put at the end so it's what you remember.

Your folder will not look exactly like mine. Your entry file will be worded differently, your template will have its own accent, your `soul.md` will be yours or won't exist. And every bit of that is fine, because, and this is the mindset this chapter has been sneaking up on, there is no correct AI-OS. There's no exact right wording for these files, no magic phrasing to hunt down, no configuration to copy from a forum. People constantly ask me for my exact files, and I always decline, not because they're secret but because they'd fit you like my shoes. The system is a conversation between you and your AI, and the files are just where the conversation settles.

What actually matters is a list short enough to memorize, and here it is, the non-negotiables: an entry file that points to a workflows directory, kept lean. One folder per workflow, one task per workflow. A `context.md` and a `logs.md` in each. The three rules written down where the AI will read them: follow exactly; stop and ask; use code to accomplish workflow tasks. Secrets out of the folder. Hold those, and every other detail is yours to shape, or to let your AI shape, in whatever words come naturally.

Everything on that list you've now seen with the panel open. You know your machine.

But a list of rules on a page is one thing, and rules that hold up under a distracted Tuesday are another. Why those particular rules? What actually goes wrong when systems skip them, and how do a few sentences in a text file keep a powerful, improvisational AI on the rails, most of the time? That "most of the time" is doing honest work in this sentence, and the next chapter is about all of it: the small set of laws that keep an AI Operating System boring, in the way that word is a compliment.

** Note: want to watch me build my AI-OS from scratch? Join SectionZero.ai where I've got video courses walking you through it step by step.

# CHAPTER TWELVE

## The Rules That Keep It Working

Aviation has a grim, useful saying: every rule in the book is written on someone's bad day. Nobody sat in a room inventing checklist items for fun. Each line exists because, somewhere, sometime, its absence cost something.

The rules in this chapter are like that, smaller stakes, same principle. Every one of them is written on a bad afternoon: a workflow that silently did the wrong thing for three weeks, a "helpful" improvisation that broke a page, an error that vanished without a trace, an instructions file that grew until the AI started ignoring the middle of it. I've had those afternoons so you can skip them.

You've already met every rule on this list; they've been woven through ten chapters and written into your folder. What you haven't seen is why each one earns its place, what actually happens when it's missing, and how a sentence in a text file manages to govern a machine that, as Chapter 3 established, has a mind of its own and a bias toward helpfulness. That last question is the real subject of this chapter, and I'll deal with it honestly, including the part most books would skip: these rules work most of the time, and you should understand exactly what "most" means and what to do about the rest.

Five rules. Here they are with their bad afternoons attached.

## Rule one: follow the workflow exactly. Don't improvise.

The bad afternoon: my page workflow, months in, running flawlessly, until a routine model update made the orchestrator a bit more creative. One day it decided, mid-run and unannounced, that my Q&A sections would engage readers better in a livelier format. It restructured them. The pages published. Nothing crashed, no error fired, and I didn't notice for days, because everything looked fine unless you knew what it was supposed to look like. Downstream, the formatting my sites depended on quietly broke.

That's the shape of the AI-native failure: not wrong, exactly, but different. Chapter 3 named the cause, stronger models drift toward what they think you really want. In a chat that's delightful. In a workflow it's a slow leak.

The rule, written in both the entry file and every workflow's `context.md`: follow the instructions exactly as written. Don't improvise, don't optimize, don't be helpful. If you see a better way, say so, and wait.

That last clause matters, and it's the part people leave out. You don't want a sullen orchestrator; its ideas are often genuinely good. The Q&A restructure might have been an improvement! The rule doesn't forbid the idea; it forbids the unilateral action. "Propose, don't impose" keeps all the intelligence and removes all the surprise: suggestions arrive as suggestions, you say yes or no, and if yes, the change gets written into the instructions where it becomes the new law, visible, permanent, and yours. Initiative flows through the same gate everything else does: the conversation.

Here's the good version, so you know it when you see it. At the end of a run: "Done and logged. One observation, not applied: the last several summaries have all flagged the same supplier's price drift, and a two-line trend note at the top would save you re-noticing it weekly. Want me to add that to the format?" Everything about that is right: the run itself came out exactly per spec, the idea arrived clearly labeled as an idea, and the decision came to you. You'll say yes to most of these, they're usually good, and every yes makes the workflow more yours. What you're training, between rule one and this, isn't obedience for its own sake. It's the difference between an assistant with initiative and an assistant with permission, and any owner who's employed humans knows those are different job descriptions.

### Rule two: on any error, stop and ask. Never work around it.

The bad afternoon, a friend's this time, with a system built before he knew this rule: his research workflow hit a source that had changed its layout. The fetch came back garbled. An eager orchestrator, wanting to deliver, did its best with the garble, filled the gaps plausibly, and produced a digest that read beautifully and was partly fiction. He acted on it.

An error that stops you costs minutes. An error that gets papered over costs trust, and sometimes money, because you don't know which outputs to doubt. The worst version isn't the workflow that fails; it's the workflow that fails and delivers anyway.

The rule: when anything goes wrong, anything unexpected, missing, malformed, or ambiguous, the orchestrator halts, writes what happened to the log, and asks you. It does not retry creatively. It does not substitute. It does not guess.

Be generous about what counts as "wrong," too, because the obvious errors, a program crashing, a file missing, were never the dangerous ones; they announce themselves. The dangerous ones are the almosts: the export that opened fine but has half the usual rows; the fetched page that loaded but looks nothing like it used to; the number that parsed correctly and is fifty times its historical range. A workflow instructed only to stop on failure will sail through every almost, because technically nothing failed.

So the instruction worth writing, and worth saying in your build conversations, is broader: stop when something is unexpected, not just broken. You'll occasionally get a false alarm, the workflow pausing to ask about a genuinely fine oddity, and you should welcome every one, the way you'd welcome a new employee asking "this invoice looks strange, is it?" That's not friction. That's the judgment you hired. You saw the good version in Chapter 6: "The transcript didn't come back. Here's what I got. How do you want to proceed?" That sentence is the sound of a system that would rather be late than wrong, and late-but-honest is the only failure mode you can build a business on.

And when it stops? Chapter 10 gave you the whole playbook: "here's the log, figure out what changed, propose a fix." Stopping isn't the breakdown; it's the handoff.

### Rule three: log every run. Start and finish, every time.

The bad afternoon: early on, a workflow of mine simply hadn't run for eleven days. Or had it? Nobody could say, which was precisely the problem. No log, no history; no history, no difference between "it's been fine" and "it's been silently dead since the 4th." Debugging without logs is archaeology without artifacts: all theory.

The rule costs one line at the start of a run and one at the end, written to the workflow's `logs.md`. Started at this time. Finished at this time, here's what happened, or, here's what went wrong. The discipline is trivial, and it's baked into your template so neither you nor the AI ever thinks about it. Note the pairing, too, and why both lines matter: an end-line alone can't distinguish "never ran" from "ran and died silently," which are wildly different problems. A start-line with no end-line is the log's way of pointing at exactly where something expired mid-run. Two boring lines, and between them, every question you'll ever ask in an ugly moment already has its answer filed. What it buys is enormous. Rule two runs on it: "stop and ask" only helps if what-happened got written down for the fix conversation. Chapter 13 is about to run on it: the log is how you audit a machine whose code you can't read. Your business runs on it in small ways you'll discover: did the summary go out? When did prices last update? Ask the log; it was there.

Together, the first three rules form the character of the machine: no improvising, no papering over, everything on the record. If that sounds less like software engineering and more like how you'd want an employee to operate, you've understood the design.

### Rule four: keep the entry file lean.

The bad afternoon is slower than the others, a rot rather than a break. You add an instruction to the entry file, just one. Next month, two more. A special case, a preference, a note about Thursdays. A year on, the file is four times its birth weight, and you start noticing oddities: instructions skipped, rules half-followed, behavior that made you say, more than once, "but it literally says right there..."

Here's the mechanism, in plain terms. Everything in that file gets loaded into the AI's working attention, every session, before any work begins. That attention is vast but not infinite, and it doesn't degrade politely: an overloaded orchestrator doesn't announce it's overwhelmed, it just gets subtly worse, and the middle of a long file is where instructions go to be forgotten. A bloated entry file doesn't merely slow things down; it actively raises the odds of the AI wandering, misremembering, or filling gaps with invention.

The rule: the entry file holds only what's global, the structure, the five rules, the pointers, and everything else pushes down into the relevant workflow's `context.md`, which is read only when that workflow runs. Detail at the point of use; a doorway, not a warehouse. When you're tempted to add a line, ask one question: is this true of every workflow? If not, it lives downstairs. Twice a year, spring cleaning: "read the entry file and propose what can be cut or moved down." The AI is a surprisingly willing editor of its own constitution.

### Rule five: hold the concepts, not a script.

The last rule is aimed at a failure of yours rather than the machine's, and I get to watch people commit it every month: the hunt for the magic words.

It starts innocently. Someone sees a system like mine working and asks for my exact files, my exact prompts, the precise sentences that make it go, so they can copy them character for character. The instinct is decent, that's how we learn recipes and chords, but it aims at exactly the wrong layer, and it produces a brittle kind of user: one who's afraid to touch anything, because they never knew why any of it worked, only that the incantation came from someone who seemed sure.

So here's the rule, and it may be the most liberating sentence in this book: there are no magic words. None. Your AI will word your entry file differently than mine, differently, in fact, than it would word it for you tomorrow, and it will work fine, because what makes the system work was never the phrasing. It's the concepts the phrasing carries: one folder; lean entry file pointing at workflows; one task per workflow; context and log in each; follow exactly; stop and ask; use code rather than AI when possible. That list, which you can now recite, is the entire load-bearing structure. Every sentence in every file is just one possible way of saying it, negotiated between you and your AI in the conversation that built it.

This is why I've never once given you an exact prompt in ten chapters of a book about directing AI, and it wasn't an oversight. Scripts expire; concepts compound. The person who memorized my files is helpless the day something shifts. The person who holds the concepts just explains what they want, again, in new words, to whatever AI is sitting in the chair. You've been becoming the second person since Chapter 1.

### The rules for you

Five rules for the machine, and fairness demands the other half of the constitution, because the system has two parties and the human one also has failure modes. Mine are four, learned the usual way.

Start every session with the bootstrap. Chapter 11 gave you the ritual, "read bootstrap.md and follow it", and the rule here is just: actually do it, every time, especially when you're in a hurry, which is exactly when skipping it costs the most. An orchestrator running without its standing orders is the single most preventable source of weirdness in this whole system, and prevention costs five words.

Change one thing at a time. When a workflow needs two improvements, make them in two rounds, run between them. Batch three changes and something comes out different, and now you don't know which change did it; the same discipline your mechanic uses, for the same reason. Corrections are cheap here, so there's no efficiency argument for bundling. Patience costs a minute; untangling costs an evening.

Read the output like a customer, not a proud parent. The failure mode of automation owners isn't neglect at first, it's the opposite: a honeymoon of careful checking, then months of skimming, then one day realizing you haven't actually read the Monday summary in a quarter, you've just forwarded it. Chapter 13 formalizes this with probation periods and spot checks; the personal rule underneath is simpler. Occasionally read your workflows' output as if you were paying for it. You are.

And never fix files by hand when you can fix them by conversation. You can open every file in the folder, and there's a temptation, once you're comfortable, to just edit a rule directly into a `context.md`. Resist it, gently. Not because you'd break anything sacred, it's your file, but because changes made through the AI get made consistently, in the right file, with the AI's understanding updated in the same stroke. Hand edits it discovers later are how a system develops folklore: rules nobody remembers making. The conversation is the changelog.

## What "most of the time" means

Now the honesty I promised, because there's a question underneath all five rules that deserves a straight answer.

These rules are sentences in text files, read by an AI that Chapter 3 told you has drift in its nature. So: does it obey?

Most of the time, yes. And I mean that precisely, not as hedging. Writing the rules down, in the entry file, again in each `context.md`, at the point of use, dramatically improves compliance; that's not superstition, it's how these systems weight what's in front of them, and it's why the rules live in two places. But I won't tell you the sentences are shackles. They're strong guardrails around a probabilistic mind, and once in a while, a model update, an odd context, a Tuesday, something tests them.

Here's why that doesn't keep me up at night: look at how the rules are arranged. They back each other. If rule one slips and the orchestrator improvises, rule three means the run is on the record and the drift gets caught. If a step fails strangely, rule two turns it into a conversation instead of a cover-up. If compliance ever feels mushy, rule four is usually why, an overstuffed entry file, and the fix is an afternoon's pruning. The system doesn't assume a perfectly obedient AI, that's the whole point, it assumes an excellent-but-drifty one, and surrounds it with structure, records, and stopping points, so that when drift happens, it's small, visible, and correctable. Engineers call this defense in depth. You can call it the reason a probabilistic machine can be trusted with deterministic work.

Which raises the question this whole part of the book has been circling, the one a sharp skeptic, maybe your business partner, maybe the voice in your own head, will eventually ask out loud: "Let me get this straight. You can't read the code. The brain in the middle is admittedly drifty. And you trust this with the business?"

Yes. Confidently, and for reasons I can defend to any skeptic at any dinner table, because trust was never about reading the code. The next chapter is that defense, and by the end of it, it'll be yours.

# CHAPTER THIRTEEN

## Trusting What You Can't Read

This morning you trusted your life to code you can't read.

If you drove, code ran your brakes; anti-lock braking has been software for decades. If you took an elevator, code decided when the doors could close. Your bank moved your money with code, your pharmacist's system checked your prescription against your other one with code, and if you flew anywhere recently, code flew the plane for most of the trip while the pilots supervised.

You didn't read a line of any of it. Neither did I. Neither, and this is the part people forget, did the pilot, the pharmacist, or the bank teller. Almost nobody who relies on software has read it, and that stops exactly no one, because somewhere along the way, civilization worked out how to trust what it can't inspect. Not naively. Systematically.

I open there because of a worry that surfaces around this point in the journey, sometimes in your own head, sometimes from a spouse or business partner watching over your shoulder: "You're running your business on programs an AI wrote, and you can't even read them?" It sounds devastating. It's actually a door into the most useful thing this chapter can teach you, because the moment you examine how trust in the unreadable actually works, everywhere, always, you find it was never about reading. And the system you've built in this book turns out to be unusually well equipped for how it does work.

## How trust actually works

Take the pharmacist. You hand a stranger a slip of paper and swallow what they give you. On what basis?

Not inspection; you can't check their chemistry. Your trust is built from other materials. Track record: they've filled your prescriptions before, correctly. Records: every pill dispensed is logged, batch-numbered, auditable; mistakes can't hide. Predictability: the same prescription produces the same bottle every time; pharmacies are engineered to be boring. And checkpoints where stakes spike: the pharmacist comes out from behind the counter to counsel you on the new blood thinner, not on your usual refill.

Track record, records, predictability, checkpoints at the stakes. That's the whole architecture of rational trust in things we can't personally verify, your mechanic, your accountant, the elevator, the plane. Nowhere on the list: "I read the source code." Now watch how precisely your AI-OS maps onto all four, because none of it is coincidence. The system was designed backward from this list.

## The four instruments

Predictability came first, back in Chapter 3, and it's the foundation: most of your workflow is deterministic code, and deterministic code doesn't have days. The program that parses your Monday spreadsheet parses it identically whether you read the code or not; run it on the same file twice and you get the same answer twice, this week, next year. This is a kind of trust that talent can't offer, only sameness can, and it's why the book pushed the bulk of every task into the boring 60. Here's what that means practically, and it's stronger than it first sounds: you can verify behavior instead of inspecting internals. You can't read the program (unless you're a coder; then by all means have at it), but you can feed it a spreadsheet where you already know the answer and check what comes out, and with a deterministic machine, unlike with a person or a model, passing that test once means something durable. Which is exactly what you did, without ceremony, in Chapter 10. Those first supervised runs and corrections? That was a verification protocol. You've been auditing since your first afternoon.

Records are the logs, and by now they need no introduction, only a reframe. Rule three's `logs.md`, started, finished, here's what happened, reads like bookkeeping until the day someone asks the pharmacist's question: "Are you sure it went out correctly last Monday?" Then it's the difference between "I believe so" and opening a file: ran at 9:04, read 412 rows, sent at 9:06. Systems that keep receipts get trusted; systems that ask you to take their word don't. Yours keeps receipts by constitution. And notice, from Chapter 7, that the receipts go deeper than run history: everything your system knows about you sits in files you can open and read. The mind of the machine is text. When people call AI a black box, they're right about the model, and only the model; the system around yours is glass.

Track record accumulates on its own, but you can accelerate it, and this is the most practical advice in the chapter: run new workflows in a probation period, exactly the way you'd break in a new hire. For the first stretch, the workflow does the work and you spot-check the output, not every line, just the parts you'd have gotten wrong. New payroll clerk, you check their first few weeks' math; new workflow, same instinct. Every clean run is a data point, and the data points compound fast, most workflows graduate in a few weeks, because remember what you're actually probing: the 60 is deterministic, so what you're really watching is the thin judgment slice, one small window rather than a whole machine.

Here's what a probation actually looks like on a calendar, using the Monday summary. Weeks one and two: full check. You run the workflow, then skim the raw export yourself the old way, five extra minutes, and confirm the summary said what the data says. Weeks three and four: spot check. You stop reading the raw data and instead verify one thing per week, recompute one total, sanity-check one flagged item. Weeks five and six: exception check. You read the summary like a customer and only dig when something smells off. After that: graduated. You've personally verified the machine a dozen ways at declining cost, and your confidence is made of observations, not hope. Write the graduation somewhere if you're feeling ceremonial; some people do, and I understand the impulse. It's the first employee they've ever trained who kept every single correction.

And calibrate probation to stakes, not to a standard schedule. The research digest, where a bad output costs me a shrug, got days of this. A workflow touching anything that leaves the building gets the full treatment. There's no virtue in auditing a grocery-list workflow like a bank, and no excuse for the reverse.

One more trust source deserves a paragraph, because it quietly carries more weight than people expect: the code in your folder is boring, and boring is a credential. When the AI writes a program to parse a spreadsheet or fetch a page, it isn't inventing exotic machinery; it's producing the same standard, well-worn patterns that millions of working programs already use, the stuff programming's textbooks and libraries were built around. AI models are at their most reliable exactly here, in small, conventional, single-purpose programs, the kind your workflows are made of, and at their least reliable in the sprawling, novel, interlocking systems your workflows deliberately aren't. You didn't just get code you can't understand. You got the most-tested genre of code there is, doing the least ambitious jobs it's ever been given.

And checkpoints, the pharmacist stepping out from behind the counter, you get to design in wherever the stakes deserve one. This is the instrument most people don't realize they're allowed to use, so let me be loud about it: a workflow doesn't have to end in an action. It can end in a proposal. My page builder publishes on its own, because a formatting flaw on a web page is a cheap mistake, but a workflow that emails clients? Build it to draft, not send: everything automated except the last click, which is yours. Money moves, contracts, anything customer-facing, put a human gate at the moment of consequence. You're not diluting the automation, drafting was 95 percent of the work; you're pricing the checkpoint to the stakes, precisely how every trustworthy institution you deal with already operates. Your risk tolerance, your gates, adjusted anytime; workflows earn autonomy the same way people do.

Since checkpoint design is a skill, one worked example beyond draft-not-send. Suppose you build the invoice chaser from Chapter 9's lightning round: it knows who's late and drafts the escalating nudges. Gate it in layers matched to consequence. The reading and sorting, fully automatic; being wrong about nothing, costing nothing. The friendly first reminder, automatic after probation if you like; low stakes, high volume, the whole point of the workflow. The firm second notice, draft-not-send; tone with a customer is worth a fifteen-second read. And the final "we're pausing your account" letter, never automated at all; the workflow's job is to tell you it's time and hand you the file. One workflow, four rungs, from full autopilot to advisory-only, all specified by you in plain sentences during the build. Nobody who sees that design ever asks the "how can you trust it?" question again, because the answer is visible in the architecture: trust isn't one decision, it's a dial, and you're holding it.

Four instruments, same as the pharmacy: predictability from determinism, records from the logs, track record from probation, checkpoints where consequences live. So when the skeptical partner asks the dinner-table question, here's the honest answer, and notice it never mentions reading code: "The mechanical parts behave the same every run, and I tested them against results I knew. Every run is logged, and I can show you the log. It ran supervised for a month before I let it run alone. And nothing important leaves without my click. What's your bookkeeper's audit trail look like?"

## When trust breaks anyway

A chapter on trust that didn't discuss betrayal would be selling something, so: things will occasionally go wrong. The 10 percent is probabilistic; Chapter 12 told you what "most of the time" means. What matters, for trust, is what going-wrong looks like inside this architecture, versus the alternatives.

A failure in the 60 announces itself: something changed in the world, the program errors, rule two halts the run, the log holds the details, and the fix is Chapter 10's conversation: here's the log, propose a repair. Loud, located, cheap; barely a trust event.

The quiet failure lives where drift lives, in the judgment slice: a summary that mischaracterizes, a flagged Thursday that shouldn't have been, tone that misses. Two honest things to say about it. First, the blast radius is small by design: drift can only touch what the AI touches, and after twelve chapters of pushing work into code and files, that's one thin layer, usually the prose. The numbers came from the deterministic part; the "weird Thursday" is an opinion about correct arithmetic, not wrong arithmetic wearing a disguise. Second, your checkpoints were placed for exactly this: the outputs where a mischaracterization could really cost you are the ones you gated with a human click. You read the client email before it sends. That's the system working, not failing.

So when you catch a quiet miss, and occasionally you will, the response isn't the sinking feeling of betrayed magic. It's a correction, Chapter 10 style: say what was wrong, watch it get written into the instructions, permanently. Trust in this system was never "it will never err." It's "errors are small, visible, stopped at the gates, and each one makes the system better." No employee, no vendor, no software, no pharmacist offers you a better deal than that.

## The absurd standard

Step back with me, because there's a comparison this whole discussion has been dancing around, and naming it is the chapter's real ending.

The manual process this book replaces was never audited like this. When a person built my video pages by hand, five hours each, skilled and careful, there was no log of every step. No determinism; tired Fridays produced different pages than fresh Mondays. No verification protocol; I spot-checked when I remembered to. My trust was a warm feeling built on a good interview and a few clean months, and it was probably justified, but I couldn't have defended it with evidence, and nobody ever asked me to. That's the punchline: nobody demands proof from the old way. The spreadsheet you've trusted for years, who audited that formula, the one typo'd in 2019? The "can you really trust it?" question gets aimed exclusively at the new thing, while the incumbent glides on familiarity.

You, meanwhile, are now holding the most auditable operation you've ever run: every run on the record, most of the machine incapable of moods, tested against known answers, gated at the stakes, correctable in sentences with corrections that stick. Held to the standard we hold everything else to, behavior, records, track record, checkpoints, your folder isn't the thing that deserves suspicion. It's the first part of your business that can actually survive the question.

So trust it, in the exact, earned, unsentimental way you trust your pharmacist. And then, having earned real confidence in the machine, guard against the one thing confidence does to people: overreach. Because the last skill this book has to teach you is the one that separates the operators from the tinkerers, and it runs against everything a book about automation is supposed to say.

Knowing when not to.

# CHAPTER FOURTEEN

## When Not to Automate

The most dangerous moment in your automation life is about six weeks from now.

Here's how it goes. Your first workflow works. Then your second, faster to build than the first, and your third, faster still. The rhythm is in your hands now, the machine does what you say, and somewhere in week six you catch yourself looking at your calendar the way a man with a new chainsaw looks at trees. Everything looks like a workflow. Why is anyone doing anything by hand? Why am I?

I've watched this moment arrive for a lot of people, and it arrived for me as well, so believe me when I say the following with affection: the skill that separates people who run automated businesses from people who tinker with automation is not building. Building, you now know, is the easy part. The skill is judgment about what deserves to be built, and every experienced operator I know carries a private list of things they've learned to leave alone.

This chapter is my list. Five kinds of tasks that don't belong in your workflows folder, the honest failure modes to expect from the ones that do, and a four-question test that compresses the whole thing into something you can run in your head in ten seconds. It's the shortest important chapter in the book, and it exists because a book that only told you when to say yes would be an advertisement, not an operating manual.

## Don't automate the rare

The math nobody does in week six: a workflow costs an afternoon to build, plus corrections, plus a little upkeep when the world shifts under it. That price is trivial against a weekly chore, Chapter 9's arithmetic buries it, and it is absurd against a task you do twice a year.

The quarterly-ish report you assemble differently each time, the annual insurance shop-around, the once-in-a-while data cleanup: by the time these come around again, the source has changed, the format has changed, your needs have changed, and you're not running a workflow, you're rebuilding one, which is to say you've automated nothing and added maintenance. The recurring test from Chapter 1 wasn't a warm-up exercise; it was load-bearing. Frequency is what pays automation's mortgage.

For the rare task, you already have the right tool, and it isn't a workflow: it's the conversation. Open the AI, describe the weird annual thing, do it together in twenty minutes, walk away. No folder, no upkeep, no guilt. An AI-OS doesn't make ad-hoc conversation obsolete; it makes it a choice, and for rare work, it's the right choice.

The gray zone is monthly, and here's the tiebreaker I use. A monthly task earns a workflow if it's stable, same shape every month, like reconciling statements, because twelve identical runs a year will pay for an afternoon. It stays conversational if it mutates, like "prepare for the monthly partners meeting," where every month's version is really a new task wearing the old one's name. The question isn't just how often does this recur; it's how much of it recurs. A task can happen monthly while its describable core happens never.

## Don't automate the relationship

In Chapter 4's tour of my morning, one task got handled outside the workflows folder: a delicate client email, talked through by hand. That wasn't modesty. That was policy.

Some tasks look like their mechanics but are actually their humanity. The condolence note, the price increase to a loyal customer, the "we need to talk about the invoices" email, the apology. An AI can draft any of these competently, and for some of them a draft is genuinely useful. What can't be automated is the part the recipient is actually receiving, which is evidence that a person stopped and attended to them. People can increasingly smell machine-shaped warmth, and here's the asymmetry that matters: the efficiency you gain is minutes, and the cost if it's detected is the relationship. Machine-shaped warmth is worse than no warmth; it's warmth that announces you couldn't be bothered.

I watched a sharp consultant learn this one the hard way. He automated his networking follow-ups, the "great meeting you at the conference" notes, complete with a personal-detail slot the workflow filled from his contact records. Efficient, and for a while, apparently fine. Then two people got nearly identical "personal" notes and compared them at the next conference. The machinery wasn't the crime; the discovery was, because the entire value of a follow-up note is the thirty seconds of attention it proves, and proof of attention is precisely the thing you cannot manufacture at scale. He'd automated away the product and kept the packaging.

The rule of thumb: automate the paperwork around relationships, the scheduling, the records, the reminders, the follow-up logistics, and keep the relationship itself in your own hands. Your workflows should buy you the time to write the hard email yourself, not write it for you. That's not a limitation of the technology. It's what the time was for.

## Don't automate what you can't check

Every workflow you trust, you trusted through Chapter 13's instruments: you fed it inputs where you knew the answer, watched it through probation, gated the stakes. Now notice what all of that assumed: that you can tell good output from bad.

Sometimes you can't, and those are the tasks to refuse. The contract clause you couldn't evaluate yourself, the tax position you can't verify, the technical claim outside your competence: automate those and you haven't delegated a task, you've delegated your judgment, with no way to notice when it fails. Chapter 13's probation only works if the probation officer can recognize a mistake. A workflow whose output you can't check isn't an automation; it's a bet, renewed silently every run.

The related trap, subtler and more common: automating a task you haven't yet done yourself enough to describe. You'll be tempted, someday, to automate your way out of learning something, to build the workflow for the new marketing channel you've never actually worked by hand. But Chapter 10 taught you what a build conversation is made of: your description, your edge cases, your Thursdays. You can't transfer taste you don't have. Automate mastery, never mystery: earn the boring version first, then hand it over. The good news is that the earning can be quick, a few manual cycles is often enough, and doing them with the AI as your assistant is a fine way to learn faster. The workflow comes after.

## Don't automate your eyes

This one took me longest to learn, and it's the one I'd most like to save you from learning slowly.

Some manual chores are secretly surveillance. Skimming your own customer emails, eyeballing the week's numbers before they're summarized, reading a few raw reviews, walking the store, whatever your version is: these chores have a boring surface and a precious core, because they're where you notice things nobody told you to look for. The complaint phrased strangely, twice. The number that's fine but feels off. The competitor mentioned in passing. A summary, by definition, contains only what the summarizer thought mattered; the noticing happens in the residue, the stuff that would never survive summarization because nobody knew it mattered yet.

Automate all of it and the reports keep arriving, cleaner than ever, while you slowly go blind in the specific way that sinks owners: current on everything, in touch with nothing.

The design answer isn't to refuse the workflow; my digest from Chapter 9 is exactly this kind of automation, and I'd never give it up. The answer is to automate the coverage and keep a deliberate patch of rawness: my digest summarizes the many, and even so, I still read a handful of raw sources myself, on purpose, most weeks. Tiered reading, remember, was designed to let me decide how deep to go, not to make sure I never touch bottom. Keep one manual ritual in every domain you'd be sorry to lose your feel for. Call it the owner's walk. Schedule it if you have to. It will feel inefficient, and it will save you, eventually, from a surprise that no workflow could have flagged, because flagging it would have required knowing in advance what surprise was coming.

### Don't automate for sport

Last one, said plainly because week-six you needs it plain: some workflows get built because they're fun to build, not because they save anything, and building them is a leak dressed as productivity.

The tell is simple. A workflow is a tool if it retires a cost you actually pay: hours, dollars, dread, errors. It's a toy if, honestly audited, the chore it replaces cost you four minutes a month, or wasn't a chore at all. Toys aren't evil, tinkering is how you got good, and a hobby is allowed to be a hobby. But every workflow, tool or toy, is a thing that can break when the world shifts, a thing that pings you, a thing on the maintenance rolls, so a folder full of toys is a folder that costs attention and returns cleverness. Meanwhile the actually expensive chore sits unautomated, because it was less fun.

The discipline: before you build, say out loud what the workflow retires. "This saves me thirty Monday minutes." "This replaces ninety dollars a page." If the sentence comes out "this would be cool," build it on a rainy Saturday and know what it is, or don't, and spend the afternoon on the boring one worth real money. The measure of your AI-OS was never how many workflows it holds. It's what the workflows hold back.

## What breaking looks like

Even the deserving workflows, the recurring, checkable, relationship-free, genuinely valuable ones, live in a world that shifts, so here's the honest weather report, the failure modes you should simply expect as operating conditions rather than betrayals.

Sources change; fetchers break. The site redesigns, the export adds a column, and the deterministic step fails, loudly, thanks to rule two. This is the most common failure in my system, it happens a few times a year across dozens of workflows, and the fix is the standard ten-minute conversation with the log. Models update; flavor drifts. The judgment slice writes a little differently after an upgrade; tone shifts, formats wobble at the edges. Rule one contains it, probation habits catch it, and Chapter 8 is your lifeboat if an update truly misbehaves. And rarely, weirdness: the sync conflict from working two machines at once, the log that shows a run you don't remember asking for, the output that's just off. Stop, read the log, ask the AI to investigate itself; it's a surprisingly capable detective when handed its own receipts.

None of this is the system failing. This is what maintaining any real operation looks like, compressed to its minimum viable form: occasional, legible, and fixable in conversation. Budget a little attention for it, the way you budget for tire rotations, and it will never once feel like crisis.

The budgeting can even be a workflow-shaped ritual, and this is the last practice I'll hand you in this book, so take it: the quarterly review. Four times a year, sit down with your AI and walk the folder together. The conversation is nearly a script: "Read every workflow's logs since the last review. Which ones ran clean, which erred, which haven't been run at all?

Read the memory and instruction files and flag anything stale. Anything you'd propose?" Twenty minutes, and out of it falls everything this chapter cares about: the dead workflow you can retire (retirement means dragging a folder out; savor that), the toy you can admit was a toy, the drifting instructions worth pruning, and, listen for it, the task you're now doing by hand that's earned its build. The review isn't bureaucracy. It's the owner's walk, applied to the machine itself, and a system that gets walked four times a year simply doesn't develop the creeping wrongness that kills every unattended tool eventually.

## The ten-second test

Everything above compresses into four questions. Run them against any candidate before you build; they take ten seconds once they're yours.

Is it recurring? If not, just have the conversation. Could I check its work? If not, the judgment isn't ready to delegate, or isn't yours to delegate. Would a machine doing it change what it means? If yes, it's a relationship, not a task; automate around it, not through it. And what does it retire? If the answer isn't hours, dollars, or dread, it's a toy; treat it as one.

Watch the test run on three candidates, at speed. "Answer the same eleven customer questions by email": recurring, yes; checkable, completely, you know the right answers cold; meaning-changed, no, nobody wants a handcrafted store-hours reply; retires, an hour a week of tedium. Build it, probably with draft-not-send until it earns autonomy. "Write my condolence note to a client who lost her husband": the third question detonates and nothing else matters. Never. "Pick which conference to sponsor next year": recurring, barely, once a year; checkable, not really, you won't know for eighteen months; and what it retires is thinking you should be doing. Conversation, not workflow, let the AI gather and compare, and make the call yourself. Ten seconds each once the questions are yours, and notice the test never asks what's technically possible. All three are technically possible. That was never the question.

Four yeses, build it, with the rhythm from Chapter 10 and the gates from Chapter 13. Anything else, walk away happy, because saying no is not a failure of ambition. It's the ambition: the whole point of owning a machine that can do almost anything is getting to be deliberate, at last, about what it should.

And that word, deliberate, is where this book has been heading all along. The workflows were never really the destination; they're what you built along the way to becoming a particular kind of person, the one from Chapter 1's promise, the one who directs. One short chapter remains, and it's about that person.

You, six weeks from now, chainsaw in hand, choosing your trees.

# PART V: ONWARD

# CHAPTER FIFTEEN

## Becoming the Orchestrator

It's 11:40 on a Tuesday night, some months from now, and you're in bed with your phone, and there it is again: another headline, another company, another executive saying "efficiency" and "AI" in the same paragraph.

Here's what I want for you, and it's the reason this book exists: I want that headline to read differently now. Not because the world changed. Because you did.

The person who started this book read that headline as bad weather, something enormous rolling in that would do whatever it did to whoever was standing there. If this book has worked, you now read it the way a builder reads a report about power tools: with interest, maybe, but without the knot. Somewhere between the empty folder and your first "run my..." Monday, the technology in those headlines stopped being something that happens to you. It became something you point.

That shift has a name, and I've been using it since Chapter 4 without making it a job title. Let me make it one now.

You're the orchestrator. Not the AI; that was always a small sleight of hand in the metaphor, and you probably caught it chapters ago. The AI conducts the workflows, sure. But you conduct the AI, and you decide what music gets played at all, and in any orchestra that matters, the sound belongs to the person who chose the program. The AI is the best hire you'll ever make: brilliant, tireless, oddly priced, a little drifty, and completely without wants. Every want in your system is yours. Every workflow in your folder is a decision you made about what your time is for. That's the actual skill you've built, and notice how little of it is technical. You learned to describe work, to split doing from deciding, to correct like a manager instead of re-explaining like a stranger, to put gates where consequences live, and to say no to the chainsaw. Those are owner's skills. The folder is just where they became visible.

And here's the part I can promise about the years ahead, precisely because nobody can promise anything about the years ahead. The models will change; you're insulated, you watched me swap a driver in Chapter 8, and you know it takes an hour or two. The tools will be renamed, acquired, sunset; your folder doesn't care. The prices will move; your meter barely runs. The headlines will keep coming; they're about people who bet everything on the 90, and you built on the 10. Every force that makes this moment feel unstable to everyone else is a force your architecture was specifically designed to shrug at. You didn't learn a tool. Tools expire. You learned a posture toward tools, and postures compound.

So what do you actually do now, tonight, tomorrow? Three things, and none of them is "reread this book."

Build the second workflow. The first one taught you the rhythm; the second one makes it yours, and the drop in effort between them is where most people finally, viscerally believe. Then the third. Your list from Chapter 1 has plenty left on it, and your list will grow back, it always does, because noticing is a habit now.

Give one away. Somewhere near you is a person with the exact knot you had in Chapter 1, a spouse, a friend with a struggling shop, someone at work. Sit with them for an afternoon and build their first workflow with them. Partly because it's kind. Mostly because teaching this is how it becomes permanent in you, and because watching someone else laugh at their first successful run is, I'm telling you, better than your own.

And keep one hour manual, on purpose. The owner's walk from Chapter 14. The machine's whole gift is time; spend some of it being the kind of owner no machine replaces, the one who notices.

That's it. That's the whole assignment, for the rest of your working life: build, give, notice. The dull third of your week belongs to the folder now. What you do with the returned hours was never mine to assign.

I'll leave you where we started, because I owe that earlier version of you a final word. That night, phone six inches from your face, the fear underneath the scrolling was never really about AI. It was about ending up on the wrong side of a divide you couldn't name. Now you can name it. It was never coders and non-coders, young and old, technical and not. It was the people who direct and the people who wait, and the price of crossing over turned out to be one empty folder and the willingness to describe your own work out loud.

You've crossed. Put the phone down. The machine can watch the headlines for you; it'll brief you in the morning, for a nickel.

* * *

*If you want company on the road, that's what the next page is for.*

# GO DEEPER

This book taught you the system. Everything in it is complete: the concepts, the method, the build. You don't need anything else to keep going, and the best next step is the one Chapter 15 gave you: build the second workflow.

But a book is paper, and paper holds still. The models change, the tools get renamed, and new workflow patterns keep emerging. If you want to keep current, and if you'd rather watch some of this done over my shoulder than read it, that's what my community is for.

At **SectionZero.ai** you'll find the AI Operating System video courses, where I build workflows end to end on screen, including the full versions of all five workflows from Chapter 9. There's a growing library of builds that didn't fit in a book, help when you're stuck mid-build, and, maybe most valuable given everything Chapter 8 taught you, ongoing updates as the AI landscape shifts, so your system stays current long after this printing.

The book was the method. The community is the practice, kept up to date.

**SectionZero.ai**

# APPENDIX A

## SPECIAL SECTIONZERO.AI OFFER

Learning never stops, especially in AI. There's always something new and cool to learn. I've tried to build a website that makes it super easy to learn this stuff, and it's called SectionZero.AI

Each course at SectionZero.AI is a series of videos where you watch over my shoulder as I build something.

In one course I build the exact AI Operating System I describe in this book. In another I build a personal journal wiki that uses AI to transcribe my handwritten journal entries and create an updated real-time wiki of my life. In another I use AI to build a series of coding books to sell online as a side hustle.

I'm constantly adding courses and teaching new AI technologies as they emerge.

Membership is $79 per month or $195 for a lifetime membership but...

**AS A SPECIAL THANK YOU FOR READING THIS BOOK...**

I'd love to see you over at SectionZero.AI and I'd like to bribe you to join today; so I'm giving you a special $50 off lifetime membership (you pay just $145 for total lifetime access)

**It's my gift to you! https://SectionZero.ai/bookoffer**

**And I offer a month-long 100% money back guarantee. Check out the site, if it isn't for you...just shoot me a message and I'll immediately refund your money, no questions asked, no hoops to jump through.**

**HANDS ON HELP**

**Membership doesn't just get you videos...you also get hands on help from me and other members. Any time you get stuck with something from a video, you can post a question to me directly, or post a question directly to the community itself.**

**It's a great resource and I hope you'll take advantage of it.**

**http://SectionZero.ai/bookoffer**

# GLOSSARY

*Plain-English definitions, in the sense this book uses each term.*

**AEO (answer engine optimization).** The craft of making your content the source that AI answer engines (ChatGPT, Perplexity, Google's AI summaries) draw on and cite when people ask questions, the way SEO targets traditional search results.

**Agent (AI agent).** Software that uses an AI model to work toward a goal on its own, deciding step by step what to do next. Fully agentic tools route nearly every step through the model, which is why they're flexible, and why the meter runs.

**AI Operating System (AI-OS).** The system this book teaches: one folder holding your workflows, their code, data, and logs, with an AI as the thin coordinating layer on top. A way of organizing work with AI, not software you install.

**API (application programming interface).** The standard way one program talks to another service: how your workflow asks FRED for data or asks an image AI for a thumbnail.

**API key.** A long string of characters that unlocks a service through its API, like a password that never asks twice and can spend money silently. Keys live on your machine as environment variables, never inside your folder.

**Bootstrap file (`bootstrap.md`).** The small startup file at the top of an AI-OS. Your first message each session, "read bootstrap.md and follow it," guarantees the AI loads its standing orders.

**Context file (`context.md`).** A workflow's own instruction file: what the task is, the steps, the rules. The authority on how that workflow runs; detail lives here so the entry file stays lean.

**CSV.** A spreadsheet saved as plain text, the everyday export format of booking systems, point-of-sale systems, and the like. Comma-separated values, if anyone asks.

**Deterministic.** Behaving the same way every time given the same input, the defining virtue of ordinary code. Deterministic steps fail loudly and traceably; the opposite of probabilistic.

**Entry file (`CLAUDE.md` / `AGENTS.md`).** The file an AI app reads on arrival in your folder: what this place is and how to behave. Different apps look for different filenames; the contents carry over.

**Environment variable.** A named note stored with your computer's account settings rather than in any folder. The standard home for API keys, so secrets never travel when your folder syncs or gets shared.

**GitHub.** A service that keeps versioned snapshots of a folder: history, backup, sharing, and an undo button for your whole system.

**LLM (large language model).** The kind of AI this book runs on: a model trained on enormous amounts of text, rented by the token. Claude and GPT are LLMs.

**Log (`logs.md`).** A workflow's run history: started, finished, what happened, what went wrong. The receipts that make an unreadable system auditable.

**Markdown (`.md`).** A plain-text file format. Despite the technical-sounding name, a Markdown file is just text: double-click it and read it.

**Memory (state).** What a workflow needs to remember between runs, preferences, what it has seen, running data, stored in files inside the workflow's folder. Distinct from the log: memory is what it knows; the log is what it did.

**Model.** The AI itself, the rented brain. In an AI-OS, deliberately the smallest, most swappable part of the system.

**Orchestrator.** The AI in its proper role: firing code, shaping output, making the few judgment calls. The conductor, which plays no instrument. (You direct the orchestrator.)

**Principal.** You: the person the system serves. Optionally described in a `principal.md` file so the AI knows who it's working for.

**Probabilistic.** Capable of producing different outputs from the same input, the nature of AI models. Powerful for judgment, wrong for machinery, which is why the machinery is code.

**SEO (search engine optimization).** The established craft of getting pages to rank in traditional search results on Google and Bing.

**60-30-10.** This book's memory hook for proportions: roughly 60 percent of a workflow is code, 30 percent is data and memory, 10 percent is AI. A concept, not a measurement; the law it protects is "keep the AI the smallest part."

**Token.** The unit AI companies bill in: a chunk of text roughly three-quarters of a word. Everything a model reads, thinks, and writes is metered in tokens, which is why the design goal is a model that reads, thinks, and writes as little as possible.

**Workflow.** A task broken into steps that run the same way every time, living in its own folder with its instructions, code, memory, and log. One workflow, one task.

# NOTES

# NOTES

# NOTES

# NOTES

www.ingramcontent.com/pod-product-compliance
Lightning Source LLC
LaVergne TN
LVHW060036210926
848481LV00013B/3881
* 9 7 9 8 9 9 8 9 0 6 1 2 1 *